Arms and Artificial Intelligence
Weapon and Arms Control Applications of Advanced Computing

sipri

Stockholm International Peace Research Institute

SIPRI is an independent international institute for research into problems of peace and conflict, especially those of arms control and disarmament. It was established in 1966 to commemorate Sweden's 150 years of unbroken peace.

The Institute is financed mainly by the Swedish Parliament. The staff, the Governing Board and the Scientific Council are international.

The Governing Board and Scientific Council are not responsible for the views expressed in the publications of the Institute.

Governing Board

Ambassador Ernst Michanek, Chairman (Sweden)
Egon Bahr (Federal Republic of Germany)
Professor Francesco Calogero (Italy)
Dr Max Jakobson (Finland)
Professor Dr Karlheinz Lohs (German Democratic Republic)
Professor Emma Rothschild (United Kingdom)
Sir Brian Urquhart (United Kingdom)
The Director

Director

Dr Walther Stützle (Federal Republic of Germany)

sipri

Stockholm International Peace Research Institute
Pipers Väg 28, S-171 73 Solna, Sweden
Cable: PEACERESEARCH STOCKHOLM
Telephone: 46 8/55 97 00

Arms and Artificial Intelligence

Weapon and Arms Control Applications of Advanced Computing

Edited by
Allan M. Din

sipri
Stockholm International Peace Research Institute

OXFORD UNIVERSITY PRESS
1987

Oxford University Press, Walton Street, Oxford OX2 6DP
Oxford New York Toronto
Delhi Bombay Calcutta Madras Karachi
Petaling Jaya Singapore Hong Kong Tokyo
Nairobi Dar es Salaam Cape Town
Melbourne Auckland
and associated companies in
Beirut Berlin Ibadan Nicosia

Oxford is a trade mark of Oxford University Press

Published in the United States
by Oxford University Press, New York

© SIPRI 1987

All rights reserved. No part of this publication may he reproduced,
stored in a retrieval system, or transmitted, in any form or by any means,
electronic, mechanical, photocopying, recording, or otherwise, without
the prior permission of Oxford University Press

British Library Cataloguing in Publication Data
Arms and artificial intelligence: weapon and
arms control applications of advanced
computing
1. Weapons systems—Data processing
2. Artificial intelligence—Military
applications
I. Din, Allan M.   II. Stockholm
International Peace Research Institute
623.4′028′563   UF500
ISBN 0–19–829122–1

Library of Congress Cataloging in Publication Data
Data available

Set by Wyvern Typesetting, Bristol
Printed and bound in
Great Britain by Biddles Ltd
Guildford and King's Lynn

Preface

The importance of information technology in modern weapon systems is constantly growing. Since the beginning of the 1980s, artificial intelligence has become a significant part of this trend because of its potential applications in military decision-making. This book is a first attempt to present a broad overview of the prospects for information technology in general, and machine intelligence in particular, in the context of international security.

As it is often the case with a new and rapidly developing area, relevant material is quite scattered and inaccessible to non-specialists. In order to assemble some of this material, I convened and organized a SIPRI Workshop on 'Arms and Artificial Intelligence' in November 1986 at which the knowledge and experience of experts from six different countries were brought together. This book results from that meeting.

An overview of the prospects for artificial intelligence in weapon and arms control applications is presented in Part I. The workshop papers constitute Parts II, III and IV of the book and provide a technical, strategic and political analysis of the dangers and promise underlying weapon and arms control applications of computers and artificial intelligence. Part II gives an introduction to basic aspects of artificial intelligence concepts and computer hardware. In Part III, military and strategic implications are discussed, including the Strategic Computing Program of the US Defense Advanced Research Projects Agency, the automated tactical battlefield, strategic defence systems and political perspectives. Finally, in Part IV, there is a description of arms control applications of information technology and artificial intelligence, including verification, simulation, modelling and negotiation.

The help of Annika Strand and Fredrik Bergström in the workshop secretariat was highly appreciated; Gerd Hagmeyer-Gaverus provided many useful suggestions; and the competent editorial assistance of Billie Bielckus is gratefully acknowledged.

SIPRI ALLAN M. DIN
1987

Contents

Preface v

List of contributors xi

Abstracts xiii

Part I. Overview and summary

Chapter 1. The prospects for artificial intelligence in weapon and arms control applications 3
Allan M. Din

    I. The quest for artificial intelligence 3
    II. Computer highlights 8
    III. Military applications 14
    IV. Arms control prospects 20
    V. Summary and outlook 23

Part II. Artificial intelligence concepts and computer technology

Chapter 2. An introduction to artificial intelligence 33
Robert Dale

    I. Introduction 33
    II. What is AI? 33
    III. Expert systems 41
    IV. Future prospects 45

Chapter 3. Hardware requirements for artificial intelligence 47
Lennart E. Fahlén

    I. Introduction 47
    II. Symbolic versus numerical computing 47
    III. Programming languages for AI 48
    IV. Requirements for symbolic processing 51
    V. Sequential processing 56
    VI. Parallel processing 60
    VII. Developments in VLSI and related areas 68
    VIII. Concluding remarks 71

4. Comparison of human and machine intelligence in the context of conflict — 75
Alex M. Andrew

- I. Natural and machine intelligence — 75
- II. AI applications related to conflict — 76
- III. Well-defined problems — 78
- IV. Departures from the well-defined problem area — 78
- V. The Prisoner's Dilemma — 80
- VI. Linguistic analysis — 81
- VII. Getting around the Prisoner's Dilemma — 81
- Appendix 4A. The problem of the three wise men — 84

Part III. Military and strategic implications

Chapter 5. The Strategic Computing Program — 87
S. Ingvar Åkersten

- I. Background — 87
- II. DARPA — 88
- III. The strategic computing plan — 90
- IV. Initial SCP achievements — 95
- V. The Second Strategic Computing Program plan — 98
- VI. Afterthought — 98

Chapter 6. Artificial intelligence and the automated tactical battlefield — 100
Randolph Nikutta

- I. Introduction — 100
- II. Artificial intelligence and automation on the tactical battlefield — 100
- III. Command and control of tactical nuclear weapons — 119
- IV. Escalation risks of new war-fighting doctrines for the automated battlefield — 123
- V. Conclusion — 130

Chapter 7. Software and systems issues in strategic defence — 135
Herbert Lin

- I. Strategic defence — 135
- II. Software and systems — 136
- III. Failure modes — 143
- IV. Control — 146
- V. Artificial intelligence — 147
- VI. Deterrence versus defence; confidence versus reliability — 149
- VII. Conclusion — 150

Chapter 8. Artificial intelligence and disarmament   153
Gennady B. Kochetkov, Vladimir P. Averchev and
Viktor M. Sergeev

    I. Introduction   153
    II. Technological progress and the military–political balance   154
    III. New information technology   156
    IV. Artificial intelligence and strategic stability   160

Part IV. Applications in arms control analysis

9. Computer applications in monitoring and verification technologies   165
Torleiv Orhaug

    I. Introduction   165
    II. Computer image processing   166
    III. Scene description and AI   169
    IV. Image processing and speed   171
    V. Image analysis for verification   174
    VI. Conclusion   177

Chapter 10. Knowledge-based simulation for studying issues of nuclear strategy   179
Paul K. Davis

    I. Introduction   179
    II. Breaking out of a cultural heritage   180
    III. Relatively new developments permitting new methodologies   182
    IV. A first step towards a more ambitious methodology   186

Chapter 11. Verification and stability: a game-theoretic analysis   193
Steven J. Brams and D. Marc Kilgour

    I. Introduction   193
    II. The Verification Game   194
    III. Inducement in the verification game with detection   198
    IV. Nash-equilibrium strategies   201
    V. Conclusions   204
Appendix 11A. A game-theoretic analysis   207

Chapter 12. ARMCO-1: an expert system for nuclear arms control   214
Allan M. Din

    I. Formulation of the arms control problem   214
    II. Setting up an expert system   217
    III. Conclusions and outlook   221

Index   225

List of contributors to this book

Dr Alex M. Andrew
Viable Systems
Splatt Mill
Chillaton
Lifton
Devon PL16 0JB
England

Dr Vladimir P. Averchev
Institute of US and Canadian Studies
Academy of Sciences of the USSR
2/3 Khlebny per.
Moscow
USSR

Professor Steven J. Brams
Department of Politics
New York University
25 Waverly Place
New York, NY 10003
USA

Dr Robert Dale
Centre for Cognitive Science
University of Edinburgh
2 Buccleuch Terrace
Edinburgh EH9 8LW
Scotland

Dr Paul K. Davis
Rand Strategy Assessment Center
The RAND Corporation
1700 Main Street
PO Box 2138
Santa Monica, CA 90406
USA

Dr Allan M. Din
Stockholm International Peace Research
  Institute
Pipers Väg 28
S-171 73 Solna
Sweden

Mr Lennart E. Fahlén
Swedish Institute for Computer Science
Box 1263
S-163 13 Spånga
Sweden

Professor D. Marc Kilgour
Department of Mathematics
Wilfred Laurier University
Waterloo
Ontario N2L 3C5
Canada

Dr Gennady B. Kochetkov
Institute of US and Canadian Studies
Academy of Sciences of the USSR
2/3 Khlebny per.
Moscow
USSR

Dr Herbert Lin
Center for International Studies
Massachusetts Institute of Technology
Cambridge, MA 02139
USA

Dr Randolph Nikutta
Berghof-Stiftung für Konfliktforschung
Alkensteinstr. 48a
1000 Berlin 33
Federal Republic of Germany

Dr Torleiv Orhaug
Försvarets Forskningsanstalt
Box 1165
S-581 11 Linköping
Sweden

Dr Viktor M. Sergeev
Institute of US and Canadian Studies
Academy of Sciences of the USSR
2/3 Khlebny per.
Moscow
USSR

Dr S. Ingvar Åkersten
Försvarets Materialverk
Banérgatan 62
S-115 88 Stockholm
Sweden

Arms and Artificial Intelligence
Edited by Allan M. Din
Oxford University Press, Oxford, 1987, 243pp.
(Stockholm International Peace Research Institute)

ISBN 0–19–8291221

Abstracts

DIN, A. M., *The prospects for artificial intelligence in weapon and arms control applications.*
This paper gives a concise overview of some basic features of information technology and artificial intelligence which are relevant for weapons and arms control applications. A discussion of general conceptual and practical aspects of the technology is followed by a description of specific hardware and software developments. Finally the implications for weapon systems, arms control verification and modelling are analysed with special emphasis on the effects on decision-making processes.

DALE, R., *An introduction to artificial intelligence.*
The research area of artificial intelligence is described in terms of its programming languages, like Lisp and Prolog, and of various techniques, such as search and knowledge representation. A number of application areas are discussed with particular mention of some important expert-system techniques and their practical relevance.

FAHLÉN, L. E., *Hardware requirements for artificial intelligence.*
A detailed account is given of the computer-hardware requirements for symbolic processing versus traditional numerical computing. Various technical features, such as memory and stack management, are discussed and the problems and prospects for new techniques, in particular parallel processing, are analysed. Some of the computers either available or under development, which are dedicated specifically to artificial intelligence applications, are briefly described.

ANDREW, A. M., *Comparison of human and machine intelligence in the context of conflict.*
The significance of the field of study denoted as artificial intelligence is discussed, and it is argued that machine intelligence is unlikely to rival human performance in certain application areas. In these, the human is able to use himself as a model of other intelligent entities in the environment. The point is amplified by reference to the Prisoner's Dilemma model for international tension.

ÅKERSTEN, I. S., *The Strategic Computing Program.*
An overview is presented of objectives and accomplishments of the Strategic Computing Program initiated by the US Defense Advanced Research Projects Agency in 1983. The analysis is based on the original *Strategic Computing* document and a more recent planning document as well as on recent published accounts.

NIKUTTA, A. R., *Artificial intelligence and the automated tactical battlefield.*
Many new developments in military technology are signalling increasing automation of tactical warfare, and artificial intelligence is becoming an important element in a number of planned weapon systems. A detailed analysis is given of the implications of these trends for decision-making, command and control, and crisis stability in the context of new war-fighting doctrines in Europe.

LIN, H., *Software and systems issues in strategic defence.*
A ballistic missile defence system is likely to be integrated with other elements of strategic defence such as anti-satellite and air defence systems. The coupling of these different systems is the most likely source of accidental nuclear war or inadvertent escalation. Computers will play an essential role in system control and failures could arise when pre-programmed software has to function in an environment unanticipated by system analysts.

Artificial intelligence techniques are not likely to improve the ability of humans to analyse system requirements.

KOCHETKOV, G. B., AVERCHEV, V. P. and SERGEEV, V. M., *Artificial intelligence and disarmament.*
New information technology and artificial intelligence are becoming an important factor in advanced weapon systems which may transform the military-political situation of the world. It is argued that such developments have a negative effect on arms control and that shortened decision times and unreliable technology could lead to crisis instability and loss of control.

ORHAUG, T., *Computer applications in monitoring and verification technologies.*
The state of the art of computerized image processing is reviewed and the fundamental difficulties of scene analysis and description are pointed out. The role of recent artificial intelligence research in scene analysis and the image processing requirements for a satellite monitoring agency are discussed. It is concluded that many processing tasks can be carried out almost automatically, while the final image analysis adapted to verification still requires considerable contribution by trained photo-interpreters and domain experts.

DAVIS, P., *Knowledge-based simulation for studying issues of nuclear strategy.*
The thesis of the paper is that it is now becoming possible to greatly improve the way people study certain issues of nuclear strategy—notably deterrence, escalation and war termination. The new and experimental approach has three major elements: (a) a war-gaming style analysis, (b) multiscenario, game-structured simulation of global crisis and conflict, and (c) normal and intentionally human-like, rule-based models of national- and military-level decision-making. The latter two depend significantly on concepts and techniques from AI research.

BRAMS, S. J. and KILGOUR, D. M., *Verification and stability: a game-theoretic analysis.*
A two-person, variable-sum 'verification-game' between an inspector and an inspectee is analysed. The inspectee most wants its claimed compliance with an arms control treaty accepted, whereas the inspector most desires actual compliance. An optimal inducement strategy of the inspector, whereby it announces a mixed strategy to induce an optimal pure-strategy best response from the inspectee, is found. The appropriateness of inducement and other equilibrium concepts is considered in the context of superpower conflict.

DIN, A. M., *ARMCO-1: An expert system for nuclear arms control.*
It is suggested that arms control of strategic and theatre nuclear weapons may be approached in a novel analytical way by using knowledge-engineering techniques. An outline is presented of how to incorporate relevant weapon data, doctrines, perceptions and strategic assessments in an expert system which might become a useful tool for analysts and negotiators.

Part I. Overview and summary

Chapter 1. The prospects for artificial intelligence in weapon and arms control applications

ALLAN M. DIN

I. The quest for artificial intelligence

Defining the issue

From a military and arms control viewpoint, what is the reason for the current interest in a topic as seemingly esoteric as artificial intelligence? The answer is simple: artificial intelligence is the tip of an iceberg representing a huge body of information technology which is becoming a pivotal element in many key areas of importance to international security, command and control, strategic defence and verification, to name but a few. Most of this technology is new and many of its consequences, either beneficial or potentially dangerous, still await detailed assessment. Impact assessments of information technology are important in many areas of society, but there is little doubt that they are particularly critical and urgent in the domain of international security.

Some notions elude a simple, unambiguous definition and, consequently, appear to mean quite different things to different people. Artificial intelligence is a good example of such a notion which, since it gained widespread attention in the 1950s, has given rise to high hopes as well as profound misconceptions. In the 1980s, AI (the standard abbreviation for artificial intelligence) has become a paradigm which is central to a wide variety of time- and money-consuming efforts, both civilian and military, around the world. This circumstance, and others as well, gives good reason for interest in gauging the current and future impact of AI.

As it is most commonly perceived, the issue of AI is a rather confusing one because it is related to such a wide spectrum of problems and activities. An easy way out of the confusion would of course be just to define AI as some narrow area of computer science dealing with symbolic processing and special-purpose languages like Lisp and Prolog. However, it seems more appropriate to address a much wider definition whereby AI is loosely related to the amalgam of computer developments and applications within the upcoming fifth generation of computers.[1] In such a wide definition of the issue, some applications of advanced computing may appear to be more related to rapid data handling and automation than to intelligence and decision-making. They must nevertheless be considered to be part of the picture if, to a significant degree, they pertain to tasks which used to be carried out through skilled human labour.

The fundamental questions about AI concern the extent to which machines such as computers and robots can substitute for human thought and action and, further, the problems which such a substitution might create.[2] Trying to disentangle the human thought process invariably leads to very basic questions in cognitive science[3] and mathematics.[4] A main problem is to understand in more detail how the mind registers and encodes representations of the external world and to study how this process can be duplicated by a computer. Such a duplication, if possible, must of course be limited by the fact that there is no unique set of representations, simply because there is no unique way of thinking.

To develop computers which think like people, which is how the general direction of current AI efforts is often described,[5] would require a deep understanding of the inner workings of the human brain. Although much progress has been made in this field, for example concerning the functioning of memory, we are far from anything resembling a complete understanding and, much more so, from a computer duplicate of the brain. A more pragmatic definition of applied AI research, the one considered below, can be given in terms of improving on the human thought process[6] and of making the external performance of the computer more like human behaviour.[7]

One of the great pioneers in computer science, Alan Turing, had an early and very imaginative interest in AI.[8] Among other things, he was responsible for formulating the so-called 'imitation game', nowadays known as the 'Turing test', for assessing the performance characteristics of humans and computers alike. In a remarkable paper published in 1950, he starts by posing the question 'Can machines think?' and proceeds to develop a game in which a human being communicates (most appropriately by written messages) with some hidden interlocutors.[9] Through carefully posed questions and subsequent analysis of the given answers, the goal is to identify the interlocutors at the other end of the line—man, woman, or perhaps computer.

Turing concludes his paper by stating 'The original question "Can machines think?" I believe to be too meaningless to deserve discussion. Nevertheless, I believe that at the end of the century the use of words and general educated opinion will have altered so much that one will be able to speak of machines thinking without expecting to be contradicted.' Whether this conjecture will turn out to be true remains to be seen. One problem, when making such conjectures, is that as technology develops and performance improves, people's appreciation may change in a more demanding direction. For example, when pocket calculators were introduced in the early 1970s many people were quite impressed with the 'mental' capabilities of machines. The reaction would have been greater had calculators suddenly been introduced in, say, the 1920s. Today, however, pocket calculators are commonplace and inspire little awe. Similarly, industrial robots are accepted as a fact, and the human labour which they replace is more or less forgotten.

There is a very simple way of defining the central issue of current developments in AI and advanced computing, which moreover appears to be particularly appropriate in the context of weapon and arms control applications: namely, in terms of the problem of decision-making. Schematically, in most relevant problems, there is almost always a large amount of information which must be processed using a certain amount of background knowledge; on the basis of this, some kind of decision must be arrived at in the form of a synthesis or a concrete action. Thus formulated, the issue is to understand the basic processes in decision-making within the area of concern and to disentangle the extent to which computers can actually work out and execute decisions of various kinds.

The importance of decision-making in military, strategic and arms control contexts is of course well-known and considerable work has gone into understanding the underlying processes.[10] It is only lately, however, that attention has been focused on the potential uses of computers in decision-making[11] and their possible applications in areas related to security.[12] As is described below, the current rapid developments in AI and advanced computing are not only interesting technical issues in themselves but are also intimately related to a number of important issues on the national and international security scene.

Civilian prospects

The recent upsurge of interest in AI has to a large extent been initiated by developments in the civilian sector and, in particular, by the announcement in 1981 that the Japanese Ministry of International Trade and Industry was to sponsor an ambitious development of the so-called fifth-generation computers.[13] About $500 million were to be spent over a decade with the purpose of developing 'intelligent' computers for the 1990s and beyond, computers which would be able to understand natural language, to learn, to associate, to make decisions and to take action.

The effort is sponsored jointly by the government and major Japanese companies active in areas related to information technology. It is, in fact, not difficult to understand why a country like Japan should develop an ambition to be a leader in this field: the lack of natural resources makes it logical to cultivate information itself as a new resource and, as a consequence, to emphasize the development of information-related, knowledge-intensive industries. It is too early to say whether the results of the fifth-generation project will turn out to correspond to the high ambitions; certainly concrete results have been appearing more slowly than anticipated by some enthusiasts, but development is continuing in a dynamic way.

Some civilian applications of computers already exist which might be called 'intelligent' even if their scope is not too ambitious. They are nevertheless of great use in industry, for example CAD (computer-aided design) and CAM (computer-aided manufacturing).[14] The growing use of robots in a great variety of industrial applications is another example of a technical evolution which points to a future of machines reducing the need or substituting for qualified human labour. These developments are generally perceived as part of general computerization, affecting society in many positive—and sometimes negative—ways, however, rather than as examples of fifth-generation computing.

At the basis of the wide-ranging computerization lies the ubiquitous computer chip, which owes its existence and development, under various guises, to both civilian and military technological efforts. The dynamism and optimism characteristic of the evolution of Silicon Valley in California, from the second half of the 1970s and continuing into the 1980s, is now being extended to the AI development and applications area. The battle for supremacy in memory-chip production has practically been won by Japan, with the USA responding, for example, by trying to preserve the lead in special-purpose microprocessors which will be essential components in dedicated AI computer hardware.

Supercomputers represent another frontier of modern computer science characterized by high competition and much wrangling over possible uses.[15] They endeavour to push number-crunching capacity to its limits by further developing successful designs; but this capacity is in fact not that much higher than can be offered on a somewhat lower technological, but much more affordable, level.[16] A new development is the use of novel computer architectures involving parallel processing as a means to increase the processing speed, which would otherwise have an upper limit imposed by the finite speed of light and the chip size. There are hopes that parallel-processing techniques may be particularly suited for real-time AI applications.

Specific civilian applications of AI which are coming into use are the so-called expert systems, which represent a rather pragmatic way of adapting the idea of 'intelligent' machines to the real world.[17] They have already proven themselves to be of great value in the most diverse domains of expert knowledge, like medicine, geology and crop management,[18] and their use is certain to spread much further when dedicated computers and programs with a higher degree of sophistication are developed.

The weapon connection

Very early on, it became apparent to the military that computers held considerable promise for a number of important defence tasks. One of the first applications of computers, when they could barely be said to exist, came in cryptography during World War II when British intelligence performed a major feat by cracking the German Enigma machine codes. The sheer number of possible code combinations presented a mind-boggling problem to the analysts, but it was solved by devising a rudimentary computer to perform sorting, guessing and elimination tasks which would have required days or years for humans to perform. One might have been tempted to call the combination of analytical and heuristic skill displayed by these machines 'intelligent' with almost as much right as when commending the virtues of some of the AI applications currently being developed.

The number-crunching capacity of computers saw an important military use during the construction work on the thermonuclear bomb in the beginning of the 1950s. Once the design idea of employing a fission trigger to initiate chain reactions in an adjacent fusion material had been proposed it was necessary to carry out substantial modelling and simulation work to prove its viability. The precise shaping of the nuclear charges and the reflective material, still one of the semi-secrets of the hydrogen bomb, could only be developed after thousands of hours of computing on the most advanced computers of the time.

The computer acquired a different role when the proliferation of nuclear warheads, bombers and missiles started to establish nuclear deterrence as a cornerstone of military and political doctrine. It became necessary to keep track of the locations of thousands of weapons and targets with great precision; at the same time warning systems were established to give advance notification and analysis of presumed attacks from the other side. All these developments exploited the computer capacity of maintaining large data bases and of handling the complex input/output operations involved in military intelligence.

The developments of nuclear weapon doctrines during the 1960s and 1970s, notably counterforce and flexible response, put the focus on the whole problem complex of C^3I (command, control, communications and intelligence), the texture of which must ultimately determine the viability of and confidence in strategic doctrines. Computers necessarily play a central role in such problems since C^3I, in essence, is concerned with the efficiency of decision-making; this is precisely where the speed of computing enters to relieve the human decision-makers of the manipulation of an abundance of numbers and other pieces of information.

Electronics has today come to play a fundamental role in both the nuclear and conventional weapon areas, with chips and computers entering everywhere in basic weapon design, in command and control and in battle management.[19] The degree of pervasiveness of computer technology may be measured by various gauges, but one of the most important is certainly the extent to which military, strategic and political decision-making is being delegated from humans to machines through this computerization process.

Since the beginning of the 1980s this question has become more relevant than ever in the computer age, because technology now permits many weapons to operate in a semi or fully automated mode; the step to whole weapon systems being automated is therefore naturally a major concern. Such concerns are particularly relevant in connection with military computing programmes,[20] whose origins lie in the 1983 project of DARPA (the US Defense Advanced Research Projects Agency), often referred to as the Strategic Computing Initiative. This project was precisely directed towards the use of advanced computing, and AI techniques in particular, to develop specific autonomous weapons and systems for battle management in complex environments where human decision-making was seen to be inadequate. During the past couple of years, it has become clear that the most far-reaching application of such automated battle management systems would be within the framework of the Strategic Defense Initiative (SDI).

To a large extent, military research and development into AI have been supported with the tactical and strategic battlefield in mind, for instance to develop concepts such as AirLand Battle and SDI. Apart from the battlefield role of AI-type computers,[21] there is a pronounced, general military interest in looking into all kinds of possible applications of AI,[22] from the lowest tactical level to the highest strategic one. At present, there are few major successfully working military AI applications (if any), but the possibility of future deployment of AI-based weapons cannot be excluded and it is therefore an important task to assess realistically their impact, strategic as well as political.

The envisaged uses of computer and AI techniques in weapon systems give rise to both scepticism and concern, for example because of the risk of control failures leading to crisis and accidental war. There are, however, also possible applications of these techniques within arms control which may have a more positive connotation; such applications may appropriately be called computer-aided arms control (CAAC).[23] The most obvious application of this kind is in terms of the use of computer technology in verification of arms control treaties. The technical means of verification, in particular verification by satellite, have become increasingly sophisticated, involving masses of sensor data and complex assessments; this makes it natural to look into the possibilities for developing computer methods, both standard and semi-automatic, which could render the verification procedures more efficient. Another interesting computer application is the development of 'intelligent' data bases, expert systems, and so on, of relevance to modelling and simulation of arms control problems.

II. Computer highlights

The hardware

The development of computing machines has roots far back in history, from Pascal, through Babbage, to von Neumann, who was the first to characterize and implement the digital-processing hardware which has dominated the computer landscape for the past 40 years. The term hardware is used here to denote all materials and mechanisms involved in processor chips, memory media, wiring, design architecture and peripheral equipment. In other words, hardware is the material and physical basis responsible for the functioning of computers; software, on the other hand, is the instructions, or codes, responsible for operation and control.

The history of modern computer technology has seen four generations of computers; today we are in the middle of the fourth generation and beginning to peer into the fifth one. Early in the computer age, the technology relied on the use of vacuum tubes as the basic switching mechanism implementing the on-off, binary technique fundamental to computers. For this reason, computers in the 1940s and 1950s were bulky, produced much heat, and were neither very robust nor very powerful; these shortcomings limited their use to a few research and military centres.

The advent of the transistor towards the end of the 1950s brought the second computer generation of smaller and more reliable systems. Still, mainframe computers, that is, the big, state-of-the-art systems, were reserved for users with a special scientific and technical background working at dedicated computer centres. The technology of integrated circuits (IC) used in third-generation computers increased performance and made computers much smaller and easier to operate and maintain; at this point, the computer started acquiring a strong presence in a wide range of civilian and military areas.

The present, and fourth, computer generation is characterized by further miniaturization as exemplified by some current, to a large extent military, projects like the VLSI (Very Large Scale Integration) and VHSIC (Very High Speed Integrated Circuits) efforts.[24] The objective is to develop very compact systems with a very large number of components and to increase speed by producing chips with physical dimensions in the sub-micron range. Today most available chips are larger than 1 micron (1 μ=one millionth of a metre), but the goal is to mass-produce chips with a size of half a micron. With chips of this size, electric signals would have smaller distances to travel and the operation time would be shorter. The physical problem of the speed-of-light barrier remains, however, that is, signal-transmission speed in electronic components is limited upwards by 300 000 km/s; while this may seem a lot, the physical constraint is almost felt already in many current applications.

To some extent it is possible to improve the speed of chip performance by using new materials, for example gallium arsenide (GaAs),[25] instead of the ubiquitous silicon. Besides the speed consideration involved in such a choice, there is also the militarily important requirement of using computer components which are sufficiently hardened to withstand radiation in an EMP (electromagnetic pulse) or otherwise rugged environment.[26] Other techniques for developing new and speedier switching devices have been considered, for example, optical techniques.[27]

It is likely, however, that the most important progress in processing speed—so crucial for real-time military applications, in particular if these are to involve AI—must come from new computer architectures. With traditional architecture, which dates back to von Neumann, the computer performs operations one by one, that is, in a strictly sequential manner; since each operation may take a fraction of a millionth of a second, this would seem to be sufficient for many applications. In some massive computing situations, however, this approach is not fast enough and a possible solution could be parallel computing,[28] the principle of which is to split up the data manipulation into smaller operations which are treated by separate processors. There are several difficulties in assuring genuine parallel processing in multi-purpose computers and it may be that only certain problems are well-suited to this technique.

The user interface is a completely different hardware aspect which is extremely important in connection with improving performance and making computers 'user friendly' and even 'intelligent'.[29] Many developments in AI techniques point to a future characterized by interfaces using natural-language understanding and other kinds of personalized interaction between computers and human operators. This interface, together with the inherent technical limitations of machine-based decision-making, will be the key determinant of the future impact of the fifth generation of computers.

The software

The performance of computer hardware is obviously directly related to the underlying technology, which for many years has been the main focus of development. However, the relation to the quality of the instructions which make the computer run—the software—is becoming a greater and greater concern.[30] This concern is particularly acute in military applications where reliability and robustness must be pushed to extreme limits. In the defence area, there are currently so many question marks about these qualitative aspects that it is quite appropriate to talk about a software crisis.[31]

The nature of the development problems in hardware and software are quite different. Hardware builds on a technology which has matured over the years and for which the effect of basic technical constraints is beginning to be felt; apart from possible developments in design architecture involving, for example, parallel processing or neural networks, hardware prospects are thus relatively predictable. With software the situation is much more uncertain; the rules governing software development and management are quite diffuse and it is not clear how far it is possible to progress, for example concerning software for AI applications. It is this uncertainty and an accompanying lack of structure and uniformity which accounts for the software crisis.

Early in the computer age, software consisted of coded instructions written in a machine language directly understandable to the computer. This language was difficult to learn and cumbersome to use, facts which accounted for the limited acceptance and impact of computer technology. Soon, however, higher-level languages started to appear, which to various degrees bridged the gap between instructions understandable to the human operator and to the machine. In a higher-level language, instructions or statements are a shorthand for several more basic instructions, for example on memory management, which the computer can execute directly.

Today, there is an abundance of computer languages, on various levels of sophistication and for different types of application. Thus, for example, Fortran is a popular language for scientific applications, Cobol for business applications and Basic for simple general-purpose applications. Many problems of incompatibility arise, since it is not possible directly to use software written in one language for applications written in another. Also, although multiple choice may be a good feature in some respects, it either implies unnecessary duplication or difficult communication. Such considerations are particularly relevant for software applications in a military environment, where standardization and inter-operability are fundamental requirements.

To counter the existing software disparity, the US Department of Defense adopted Ada as a standard computer language: in principle all militarily-related software is to be written, or rewritten, in this language which has been specially developed for maximum facility and versatility.[32] However, many questions remain about whether this effort of standardization will succeed.[33] One problem is the widespread lack of familiarity with Ada, and uncertainty as to how efficiently coding for the very large spectrum of relevant software applications may be carried out in this language. After all, most existing languages were developed for maximum efficiency in special applications. Problems arise, for example, when trying to write software in Ada for AI applications instead of using dedicated languages.[34]

Anyone who has worked with computer software, for example a word-processing program, will know about the existence of 'bugs', that is, features which make the software behave in unanticipated and sometimes very disruptive ways. Such bugs may have their origin in very small programming errors which stand in no proportion to their dramatic consequences. Once, for example, the interchange of a comma with a full stop in a Fortran program gave rise to a grave technical problem in a planetary mission. A different example, related to erroneous input, was the failure in a laser-pointing test experiment within the US strategic defence development programme caused by entering distance data in miles instead of feet.

Equally often, bugs originate in programming errors resulting from the large size of programs in which the different parts may not necessarily interact in the right way under some unanticipated conditions. Many commercial software programs have several thousand lines of code and a number of military and space applications have involved hundreds of thousand, and sometimes several million, lines. One way of trying to solve this problem of scale is to apply structured programming. Pascal and Modula are examples of such programming languages in which the software is supposed to be written in a self-explanatory way using modules, which may be checked or validated individually.

Even using structured programming techniques, the problem of software validation is a difficult one. Several methods exist for testing the reliability of computer programs, but for large sizes there are just so many possible cases of parameters and situations that exhaustive verification cannot be carried through in practice. Alternative ways of writing software have been suggested, for example, automatic programming, whereby a computer would itself operate the required code. Such a suggestion is, however, somewhat deceptive since this kind of automation would amount to nothing more than devising yet another higher-level language prone to unanticipated errors of the human programmer.

AI and expert systems

There are a number of differences between the software and hardware for AI applications and equipment for more conventional computer uses. On the hardware side, the differences are currently not so big since standard computers are quite capable of running existing AI software. There are, however, developments in processor techniques and architecture which are beginning to offer hardware which is compact, fast and specially adapted for AI applications.

On the software side there are some conceptually important differences in terms of the particular languages and techniques used for the programming. The language used for most present-day computer applications, for example involving number-crunching and data bases, have an algorithmic character. This means that, when confronted with a particular problem, the language is used to set up a string of instructions which are executed in a linear way, possibly including one or several loops, thus providing a straightforward calculation procedure and an end result.

The languages of AI, the most well-known of which are Lisp (*List P*rocessing) and Prolog (*Pro*gramming in *Log*ic), are best described as dealing with formal manipulations of objects, symbols and their relations. For example, such manipulations are involved in AI applications for phrase parsing, whereby sentences are split into their grammatical components. AI languages are in principle used in a rather unstructured way, in the sense of not following a string-like algorithmic scheme for the instructions; in a properly working program, the instructions are supposed to be invoked automatically when required, eventually resulting in a chain of logical reasoning and a final conclusion.

Just as it is difficult to define the fundamental meaning of AI, so there exists no generally applicable theory of AI which could implement such chains of logical reasoning. Making a theory of AI may be as unlikely as, for example, making a theory of civil engineering, but there does exist a large body of useful AI techniques and principles.[35] In many problems one is confronted with a large number of options to choose between—called a search space in AI terminology—and a number of AI techniques have been developed to make 'intelligent' choices. Heuristic search is an example of such a technique, in which the size of the search space is gradually reduced using some kind of user-defined 'rule-of-thumb' to cut down the number of possibilities; such techniques may be applied successfully in games like chess.

The problem with such methods is of course that very often there is no obvious way of checking whether they really provide a good solution or choice. With large AI-language programs, in particular, it may be very difficult to follow, even intuitively, how the computer is proceeding to make its inferences; thus, the non-algorithmic structure has both its advantages and disadvantages. In fact many successful AI applications[36] have some very distinctive computational components which, in some way or another, embody an algorithm with an evaluation function as booster of the overall efficiency. This explains why AI-type applications using standard computer languages can produce good results and also indicates that suitable integration of dedicated AI languages with more conventional number-crunching languages is an interesting avenue.

Most current efforts in AI applications are not based on an anticipation of or reliance on early results of abstract AI research but rather concentrate on the more pragmatic line of approach of expert systems. These represent the most probable way in which AI techniques will be able to affect the real world, both in terms of civilian and military applications.[37] Expert systems do not set out to solve too big a class of problems but rather try to isolate a particular problem area from external parameters; it may then, for example, be possible to apply specially tailored techniques so as to make the computer a useful tool in decision-making. There are many techniques available for building expert systems[38] and in each possible domain of application many approaches are currently being explored.[39]

The general feature of expert systems which appears in most knowledge-intensive problem areas is that the problem must be attacked using a basic knowledge base of relevant data and rules. Extracting all of this from the experts in the particular problem area is the task of the knowledge engineer; he must judge which of the data and rules governing the expert reasoning are essential and, most importantly, which of these may be conveniently formulated in a way understandable to a computer. This part of building an expert system, termed knowledge engineering, is without doubt the most difficult part of devising a successful application.

The knowledge base is treated by the expert-system computer in the so-called inference engine—the software—written in a particularly suited language, executing the chain of reasoning which eventually gives rise to a formulation of a reduced number of choices or to the presentation of a solution to the human operator. The machine interface is of course a very important part of an expert system; communications with the computer must be interactive, eventually using natural language, so that new questions can be formulated based on previous computer responses. In this way, the expert system may become an effective tool in analysis and assessment, which would be cumbersome and slow using traditional methods relying on either manuals or direct human expert assistance.

Expert systems must be judged on their practical merits. It is difficult to say in advance whether one or the other approach will be successful but, as experience is gathered, it will become clear which problem areas are particularly amenable to expert-system and AI techniques. In evaluating their merits, it is not only a question of whether an approach works under some ideal conditions, but also of the risks and possible negative consequences of system failure. This concern is of course particularly relevant when discussing real-time military applications.

III. Military applications

General computerization

The military market for computers in the USA and Western Europe is flourishing whereas general computerization in countries of the Warsaw Treaty Organization (WTO), including the Soviet Union, is progressing at a much slower rate. It is difficult to give precise figures for the level of computerization in defence-related applications. In the area of personal computers for office and battlefield uses, for example, a low estimate of US-government and defence-firm acquisition is around 20 000 units per year,[40] but overall procurement of computer equipment probably represents a billion dollar market.

There are two categories of computer which are of special military interest: mil-spec and TEMPEST products (in US and NATO defence terminology).[41] Mil-spec machines must be built to stringent military specifications concerning shock, vibration, humidity, electromagnetic effects and frequency of failure. In a military environment, robustness is an absolute requirement which standard computer products cannot in general fulfil and special effects, such as the electromagnetic pulse generated by nuclear weapons, must be taken into account by protective mechanisms or new types of radiation resistant material.

TEMPEST products are rugged computers—commercially available machines have been refitted in a non-standard way to meet some of the military specifications. For example, circuit boards can be mounted on a shock-absorbent chassis, and voltage-surge protection and a metal casing limiting external electromagnetic effects can be added. The price of such rugged computer equipment is perhaps 2–3 times that of the corresponding commercial product, whereas complete mil-spec equipment would cost 5–10 times as much. This is a main reason why the TEMPEST market is developing so strongly, but an additional consideration is that commercial computers can often be delivered with the latest technology much earlier than military ones.

Computers are used in a wide area of military applications; for example, the battlefield army computers are used in combat service support, manoeuvre control, air-defence control, fire-support control and intelligence/electronic warfare. Most computer systems are fielded in the latter area and electronic warfare in particular involves computer applications which draw heavily on many advanced developments, including AI techniques.[42] Encryption and other security related concerns also depend on the ability of modern rugged computers to handle efficiently large volumes of data transmission.

Most of the smart weapons currently being deployed owe their existence to advanced microelectronics.[43] The sensors of autonomous smart munitions are coupled to small computers which in real time calculate the most efficient way of engagement; robustness and compactness are of course essential for such applications, which explains many of the efforts for miniaturization of computer components. Similar criteria are relevant in space applications where small on-board computers have many uses, some of which may in the future involve expert system techniques.[44]

Weapon projects

The most ambitious attempt to bring advanced computing, and AI techniques in particular, to the point of deploying a new-generation intelligent weapon was undertaken by the US Defense Advanced Research Projects Agency (DARPA).[45] Starting in 1983 with a budget of around $600 million for an initial five-year period, this strategic computing effort is often referred to as the Strategic Computing Initiative (SCI), a term quite appropriate in view of the many important developments parallel to the Strategic Defense Initiative.

The goal of the SCI is to develop a broad technology base (both in hardware and software) including processors, architectures and, in particular, AI disciplines such as speech recognition and understanding, natural-language computer interfaces, vision-comprehension systems and advanced expert systems. The basic efforts also involve programmes to increase the availability of AI scientists and engineers and their integration into the relevant defence industries; the lack of trained personnel with the necessary technical background is one of the biggest problems in getting the SCI to show quick results.

New developments in basic technology will be needed in most of the weapon project areas. This is so, for example, in the project envisaging the construction of an autonomous land vehicle, which will combine high-speed (i.e., very near real-time) computing technology, advanced image understanding and numerical calculation power to produce a moving machine capable of autonomous navigation and tactical decision-making.[46] The vehicle will be equipped with laser range scanners and colour TV cameras feeding information to several types of computer; it is scheduled to be able to move safely along roads of various quality with speeds ranging from 3 km per hour to perhaps several tens of kilometres per hour.

The autonomous land vehicle may be seen as part of a wider military trend where greater reliance will be put on battlefield robotics.[47] With the first tests of the land vehicle in 1985, general army interest in robotics and AI applications started to increase and ideas were put forward for the development of robots in a number of areas; hopes were even expressed that such robots might represent a technological revolution which could completely change many concepts of land warfare and the army force structure. Even if such claims may be exaggerated, the dangers of a battlefield manned by robots, for example related to erroneous attacks and run-away escalation, should not be underestimated.

Another novel weapon project of the SCI is the so-called pilot's associate.[48] The modern fighter cockpit is already heavily equipped with electronics, but the new project envisages pushing the computerization much further and developing an ensemble of integrated aircraft systems which is supposed to enhance mission effectiveness. Key ingredients in the pilot's associate system will be an expert system with advanced speech input/output and real-time computer technology involving, eventually, parallel processing.

The problems with which a fighter pilot is faced involve a mass of sensor data which provides information about the mechanical state of the aircraft and about external threats and targets. Split-second decisions must be taken which somehow integrate all these impressions, a task which is very often quite overwhelming. The expert-system approach is particularly attractive here because it may reduce the number of choices working in a semi-automatic mode or, eventually, act autonomously if the time pressure is very high. How effective such approaches to automation of the fighter cockpit are in actual combat situations still remains to be seen.

A third example of the SCI weapon projects is the navy battle-management programme, which aims to develop machine-intelligence technology for US naval commanders which will enable them to be locally in charge of planning and battle management. The problem facing a big naval carrier, for example, is that defensive postures must be adopted early to counter such threats as ballistic missiles, cruise missiles and torpedos, all of which in turn must be neutralized during a series of complex combinations of events and decisions. A demonstration addressing such battle-management problems is planned for the nuclear aircraft-carrier *USS Carl Vinson* after the development of a special combat action-team expert system and a spatial-data management system. An important factor, which must be taken into account in assessing effectiveness, is the possible failure of electronics and sensors in a complex battle environment involving nuclear weapons and EMP effects.

Expert-system techniques are also intended to play a major role in the battle-management system of NATO's new war-fighting doctrines in Europe of which the AirLand Battle concept is an important exponent. An expert system is under development which will support corps and division staffs in battle planning by, for example, analysing troop movements and planning fire support. Emerging technologies are increasingly shaping the modern tactical battlefield.[49] AirLand Battle and forward defence being prominent illustrations of this fact—concomitantly the time-scales for battlefield decisions are compressed. The practical viability of new weapon developments therefore depends less on the individual characteristics of the weapons than on whether the battle-management systems are really capable of handling the extremely rapid flow of events and decisions. Ultimately, the underlying information-technology systems could be required to act autonomously in many, if not most, situations; should this become the case also for the overall management, the perspective of a truly automated battlefield starts to materialize.

Command and control

The complexity of modern nuclear and conventional forces, strategic as well as tactical, is such that inherent vulnerabilities of command and control systems are almost unavoidable and of constant great concern to the superpowers.[50]

Their preoccupation with questions of vulnerability, reliability, survivability and endurance is reflected in continual and very costly modernizations of existing C^3I installations.

Possible threats to command and control systems include collateral nuclear weapon effects, such as the electromagnetic pulse, and attacks on satellite assets, which could disrupt both launch control centres and global military communications. C^3I modernization and upgrading therefore involve a great amount of 'radiation hardening' of existing facilities and deployment of new, redundant communication systems. Nevertheless, the C^3I systems depend crucially on high technology with computers and other electronics as essential components and problems of vulnerability are likely to persist. They may become even more acute as more sophisticated technologies, for example built on AI applications, are introduced.

A major command and control problem is decision-making under severe time pressure since this always involves a risk of precocious and/or incorrect action. For example, in case of execution of the SIOP (Single Integrated Operational Plan) of the USA in response to a nuclear attack, there would be a tendency to act promptly while command and control systems are still intact; that is, a pre-planned action must be decided upon early without possibility of later modification. The normal remedy for such a situation is to make the systems more enduring, such that crucial decisions may be postponed as long as possible. A different option, offered by advanced computer and AI techniques and implicit in the Strategic Computing Initiative, is to substitute the algorithmic execution of decisions by a more flexible action whereby, for example, retaliatory responses might be automatically executed based on pre-programmed strategic guidelines as well as on incoming data concerning threat and damage assessment. Whether options of this kind really can be implemented, or are even desirable, are open questions.[51]

The characteristics of command and control systems vary according to the specific level; on a global scale, for example, there is the US Worldwide Military Command and Control System (WWMCCS), where secure long-distance communication (by satellite, etc.) are required. On a more regional level, there is the NATO central region Automated Command and Control Information System (ACCIS), which in turn is connected to national systems such as the British WAVELL, the German HEROS and the French SACRA.[52]

The existence of a number of different systems in the NATO command and control structure gives rise to problems concerning standardization of the technical equipment and rational allocation of resources; for example, technical incompatibilities may be found in communication protocols, cryptology and multiplexing equipment. The current modernizations are therefore largely directed towards a co-ordination of research and development efforts which should result in increased inter-operability of both hardware and software.

The NATO doctrines of AirLand Battle and follow-on-forces attack (FOFA) have also initiated a number of new developments in command and control structures. For example, the US Army in Europe (USAREUR) is pursuing a programme to develop a Joint Surveillance Target Attack Radar System and an airborne multimode radar which will operate together with C^3 facilities located on aircraft and ground stations.[53] The above doctrines require a high degree of theatre-wide automation of command and control systems, the implications of which have been the subject of much analysis; thus, the UTACC system (Tactical Command and Control System of USAREUR) was the result of a detailed battlefield-automation architecture study.

Computer software is a most essential ingredient in the complex environment of modern C^3I systems in terms of overall cost and human efforts; in fact, the development and maintenance of software very often require more than 90 per cent of the project resources spent. This circumstance largely explains the considerable current interest in effective programming environments and novel approaches such as AI techniques to C^3 problems.[54] Software requirements have been determined by both the US Department of Defense and NATO (through the Command, Control and Data Processing Committee and the Military Command, Control and Information Systems Working Group) leading, for example, to regulations about general use of the programming language Ada.

The C^3I demands on far-reaching automation make the use of expert systems and other AI applications a virtual necessity. A typical use of AI techniques is in decision-support systems; an example is the US ARES project which will be a tool in the manoeuvre control function segment of the command and control subordinate system.[55] Several command and control projects are also concerned with real-time applications which all depend heavily on the use of AI.[56] However, many questions concerning efficiency and possible shortcomings of AI applications in command and control, for example in the context of AirLand Battle, remain unanswered.[57]

The picture of command and control systems in the USSR and the WTO countries displays many important differences from those of the US and NATO systems.[58] The emphasis in C^3I systems is on the control aspect, as imposed by a centralized decision-making structure, and automation was seen at an early stage as a means of assuring this 'top-down' control. Centralized computer facilities currently play an important role, but the slow emergence of new hardware technology seems to exclude any near- or medium-term changes in the traditional command and control structure.

The Star Wars perspective

As it happened, the Strategic Defence Initiative (or 'Star Wars') and the Strategic Computing Initiative were both launched in 1983. Initially, the Star Wars debate focused to a large extent on the possibility, or impossibility, of constructing particular, more or less exotic, types of weapon. However, it rather soon became evident that for any strategic-defence system to have more than marginal efficiency, a large number of rather formidable battle management and C^3 problems would have to be resolved.[59] These aspects of the SCI have therefore acquired an importance which goes far beyond the goals originally stated for the programme.

Electronics is a vital ingredient in SDI, both in terms of the hardware and the system architecture,[60] and progress in strategic-defence technology is therefore closely associated with progress in the VHSIC and the VLSI programmes. For the space-based element of SDI in particular, there are strong demands for compactness, performance and radiation hardness of electronic components.

A more significant concern than the availability of adequate chip hardware is the uncertainty about almost all key issues concerning SDI architecture, that is, the way in which the weapons of the various defence tiers are interconnected with the ground and space segments of the control structure. This interconnection presents formidable problems in systems analysis and, in particular, much hinges on the performance of the computers in the control system. To address the feasibility of various proposals for the SDI system architecture, a so-called National Test Bed will be set up with the purpose of running elaborate computer simulations of how battle management and C^3 might work under real-time conditions.[61]

The computer system for an ambitious strategic defence involving a wide area protection capability, that is, the Star Wars I perspective, must necessarily involve many automated elements which will require extensive use of AI techniques. The necessity for automation in a global strategic-defence system is a consequence of several features of the envisaged space-based weapons. First, boost-phase defence against ballistic missiles requires very short reaction times; also laser and particle-beam weapons act almost instantaneously, which means that space assets could be attacked with practically no warning time. Second, fully centralized control of weapon platforms is excluded because of concerns with vulnerability. As a consequence, the strategic defence must be controlled to a large extent by an automated, distributed computer system. In particular, on-board computers would under some conditions be required to make an independent assessment of external threats and to activate fire controls.

Because of these and many other technical features, the possibility of ever developing an SDI computer system, working according to the ambitious specifications of Star Wars I, has been called into question.[62] The demands on practically error-free software with several million lines of computer code, notably involving substantial elements of AI techniques which moreover can never be tested under realistic conditions, are extremely difficult to satisfy. On grounds of system architecture alone, it is therefore highly doubtful whether such a system has any prospect at all for acquiring a degree of reliability satisfactory for military and, perhaps more importantly, political decision-makers.

In spite of the fact that even official reports have expressed much scepticism concerning the likelihood of the Strategic Computing Initiative, and AI techniques in particular, being able to solve the basic problems of system architecture for strategic defence,[63] extensive research and development in the area continues. The above scepticism may, however, be reflected in some recent demands for development of a limited strategic defence only, with the alleged purpose of strengthening nuclear deterrence, that is, Star Wars II. Such a different scope will obviously tend to diminish the advanced-computing requirements on the system.

IV. Arms control prospects

Verification technology

Confidence in the value of arms control treaties is directly related to the effectiveness of the technologies available for treaty verification. In the current climate of suspicion between the two major military blocs, it is quite clear that no new treaties will be concluded unless their verification can be assured with sufficient confidence. The problem of precisely which type of verification technologies and which degree of confidence are required, must of course be addressed in a case-by-case manner.[64]

National Technical Means of verification have become a recurring element in the bilateral treaties between the USA and the USSR. To a large extent, they rely on sophisticated satellite-monitoring techniques, but in recent years a whole range of advanced sensor systems have been developed which offer far-reaching possibilities for verification in several arms control areas. New sensor technology has emerged thanks to recent advances in electronics and, at the same time, the computer has acquired an essential role in the processing of the vast amount of available sensor data.

Computers and electronics in general have many features which make them of great value in verification systems for arms control treaties; one example is the seismic networks for collection and evaluation of data related to underground nuclear explosions. The verification system of a comprehensive test ban treaty might involve hundreds of seismographs, some of them in unattended 'black boxes', which must communicate large volumes of data to evaluation centres by satellite. Obviously, computers are required for filtering, storing and analysing the data; on a more advanced level, the possibility of using expert systems in the data processing has been considered.[65]

The list of sensors in which modern electronic components are essential is extremely long; it includes electro-optical, infra-red, x-ray, radar and acoustic sensors. In many cases the computer processing of sensor data is rather straightforward, but in an increasing number of cases advanced processing is needed to get an appropriate output. This is so, for example, for the synthetic aperture radar (SAR) system, in which digitized radar signals are registered at different stages and subsequently assembled by a computer to present a high-resolution picture of a given area.

Satellite monitoring, as practised by the USA and USSR, already plays an important role in arms control, but new satellites and new technical approaches make this a promising field for future applications in both arms control and crisis monitoring.[66] In fact, there is now no technical obstacle in the way of establishing a satellite monitoring agency, international or regional, concerned with specific areas of arms control and crisis monitoring; in the future, lower economic cost and enhanced technical capabilities could make the idea of such an agency even more attractive.

The techniques used in acquiring image data from space are based on photographic, radar or electro-optical methods; the latter has a great capability for rapid data acquisition and analysis thanks to advanced computer processing. Outside the USA and the USSR, this has been demonstrated very recently by the French SPOT satellite in which the image data of a given scene are registered by panchromatic or multispectral sensors aboard the satellite in digital form and later transmitted to ground stations where further processing is carried out. Geometric rectification, artificial colouring, and so on, can be performed quickly by computer and pictures for various interpretation tasks become readily available.

When large areas of the earth's surface are scanned with high spatial resolution, the amounts of data increase dramatically. This calls for rapid, automatic data handling in which AI technqiues could play an important role.[67] Pattern recognition has long since been an active area of AI research and it is likely that progress in this area could be important for certain monitoring tasks of classification and interpretation. It is clear that many scene-interpretation problems will still remain so complex as to require human assistance but, nevertheless, the expert-system approach to object detection and scene description is likely to be very valuable. Independent of future AI and expert system applications, the use of new computer techniques has a great potential for increasing the speed and cutting down the cost of image processing.

Modelling and simulation

The debate on arms control problems is often dominated by rather qualitative political and strategic concepts with the only quantitative elements being a somewhat sterile enumeration of weapon stockpiles. Many serious analytical approaches to qualitative and quantitative arms control problems involving modelling and simulation have been undertaken over the years but, in view of the complexities of the real world, much remains to be understood before the processes of arms control may be assessed realistically in this way.[68]

The analytical approach has, for example, been pursued for a long time in the context of war-gaming and military strategy by various defence-related organizations. Only parts of the methodology of such modelling have been published, but it is evident that the results must have played an important role over the years in evaluating force postures and in arms procurement. In the war-gaming type of exercise characterized by a large amount of data and contingencies, the computer has been an essential tool; however, it has not been able to eliminate a number of shortcomings of the modelling approach.

One problem has been the participation of human players, which slows down the evolution of the game and limits the number of options which can be analysed in a finite time-span. A solution to such problems has been investigated using AI techniques to automate some of the decision processes;[69] whether this approach is really able to generate more reliable results still remains to be seen.

Outside this arms-modelling area, there are many other computer applications, but they have so far mostly been based on traditional approaches involving data-base management and raw calculational power. Computer simulations which are quite illustrative and helpful in understanding basic arms control issues are, in fact, not too difficult to set up.[70] The main question is, however, how to conceptualize in an adequate way the many complex parameters which determine the evolution of the arms race and the processes of arms control; there is of course no unique and simple answer, only more or less realistic approximations.

In this connection, AI techniques could have an interesting role to play. In AI applications, a first crucial step is precisely how to devise a suitable formal representation of a given problem area which, in turn, may be understood and processed by a computer; this representation may involve both numerical facts as well as rules and doctrines. As a matter of fact, conceptualization is also the first step in any human approach to problem solving, it just tends to be more implicit than the representation-building characteristic of a dedicated AI application.

Game theory offers a very interesting framework for formalizing many relevant problems in arms control;[71] even simple 2×2 games may capture essential features of the superpower arms race, deterrence and verification by assigning values to various options and outcomes and by analysing features such as equilibrium and stability. Computer analysis may be useful in the case of more complex game types and value assignments. When combined with AI-programming techniques, it seems that game theory has an interesting potential for making automated, realistic (or at the least, relevant) simulations of many arms control problems.

Arms control negotiations is another area in which modelling and simulation could play a useful role,[72] but so far analytical and computerized approaches have received little attention. One possibility is to let a computer play a direct, third-party role in negotiations by storing in its memory a range of advance negotiation positions, some of which may overlap; this would allow for an automatic and mutually acceptable compromise when the positions of the two sides fulfil certain predetermined criteria.[73]

Another interesting possibility is to develop expert systems for use by negotiators as a tool in analysis and assessment. In fact, the expert system framework appears to be well adapted to accommodate the complicated set of numerical facts, rules, doctrines and even perceptions which enter into arms control negotiations. For example, it might be possible to generate automated responses from one side to test a proposal of the other side and in this way be able to probe the viability of certain negotiation positions. The responses given by the expert system would of course not just be yes/no answers but involve details of the applied reasoning chain.

V. Summary and outlook

Most of the envisaged weapon and arms control applications of AI are ultimately concerned with the question of the extent to which information technology can handle decision-making processes in a wide variety of contexts. This question may be addressed on the following two levels, technical and utilitarian respectively:

1. Is information technology sufficiently mature to motivate planning for ambitious, large-scale applications of AI?
2. Which applications have a chance of being put to 'good' or 'bad' use respectively, as measured by the likely quality of decisions and their potential consequences?

In the context of international security, answering these questions is of importance in evaluating the technical feasibility of a number of proposed tactical- and strategic-level weapon systems and, should such systems ever be deployed, in assessing their likely effect on the arms race, crisis stability and the course of armed conflict.

Before discussing possible answers, it may be useful to make clear the relation of arms control to the potential military uses of AI and information technology. The emphasis of arms control in connection with military applications of technology is very often directed toward exposing the dangerous effects of certain weapon technologies, for example biological, chemical and nuclear arms and, subsequently, in proposing to curtail or eliminate weapons constructed on such principles. In the case of information technology, arms control must have a somewhat different emphasis; for example, even if advanced computing was used in making the thermonuclear bomb, there would have been no real point in prohibiting the construction of powerful computers to promote nuclear arms control. Similarly, AI and information technology are today so pervasive and in such general use, that there is no point in putting direct restrictions on such technology.

In fact, as section IV shows, there exists a positive approach whereby the technology may be put to good arms control use in, for example, verification. A standard arms control approach should rather be applied to the very weapons and weapon systems which have a crucial dependence on information technology, in case the analysis shows that these might give rise to developments detrimental to international security. Such an arms control effort must obviously first concentrate on identifying the weapon developments which rely heavily on the use of computers and AI techniques; this identification was sketched in section III.

Returning to the first question about the maturity of AI technology, it is clear that the high expectations associated with the AI hype during the past three or four years cannot possibly be satisfied. In the US defence industry, in particular, many exaggerated claims about 'intelligent' weapon systems have led to a fear of negative effects on government funding when it becomes clear that most of the ambitious military applications could only be expected to be fielded towards the year 2000, if ever. The more general fear that the whole AI effort, because of not being able to deliver on its promises, would somehow fade away, as happened during the 1960s and part of the 1970s, is nevertheless unfounded; there exists a pragmatic effort based on expert systems which, with less spectacular claims, is likely to lead to important near- or medium-term applications.

On the hardware-technology side, the evolution towards dedicated processors, new materials, novel architectures and futher miniaturization will make AI applications more feasible and adaptable to various environments, including military ones. Most of these developments are relatively predictable and are not likely to change the current prospects of AI hardware. Nevertheless, technological surprises based, for example, on new superconducting materials or neural networks might make the computer hardware function more effectively and, eventually, more like the human brain. Wholly new perspectives would be opened up.

Even allowing for progress in hardware, software technology would still represent a stumbling block which might seriously hamper the development of AI applications. In the military field, software quality is already a big problem in many general-purpose computer applications and these problems will only be exacerbated in connection with large-scale systems based on AI techniques. The prime example of a future problem area is the SDI battle-management system; it is difficult to imagine that such a more or less automated system, characterized by millions of lines of computer code, new programming techniques and untested large-scale features, could possibly attain a sufficiently high degree of technical reliability. The comparison with other large systems using millions of lines of code, like national telephone systems, only emphasizes the shortcomings of the technology; many technical errors do arise daily in spite of long experience and use. For a telephone system the errors represent an inconvenience which most often can be quickly removed by sending a repair team to the critical point; in a military situation involving conflict and space-based systems, such technical corrections are clearly out of the question.

The military software crisis is not only caused by simple programming or data-input errors, which come at an average of one in every thousand lines of code; the problem is of a more conceptual character since the most unexpected failures are caused by logical interconnections between different blocks of a program which, through the ignorance or omission of the human programmer, are unable to handle certain rare contingencies. Systematic methods to check programs for both bugs and, more or less intangible, conceptual errors would in most cases demand too long a testing time to be feasible. The logic programming involved in AI applications is not likely to improve this situation as compared to algorithmic programs and it will most probably become worse.

The second question, about the decisions produced using AI applications, is both related to the quality of the software and the nature of the particular application area. The latter determines if the decision times are inherently short or long and whether the decisions might have irrevocable and/or potentially disastrous effects, whereas the former is inversely proportional to how easily the application may go astray. The so-called 'no-decision-time' problems arise in extreme situations where decisions have to be taken in real time by fully automated systems; clearly, the demand on software will be particularly high since there is no possibility for a man–machine dialogue.

In the application area of military, tactical as well as strategic, command and control, it is clear that most decisions have immediate, life-or-death effects and the use of 'intelligent' machines to automate battle management therefore gives rise to several concerns. Obviously, whether the discussion is about launch-on-warning of strategic missiles or deep-strike theatre missions, no military commander or political decision-maker is, *a priori*, willing to relegate decisions to automated systems; in principle, everyone wants to retain absolute human control over military actions. The practical enforcement of such principles is, however, influenced by the availability or non-availability of systems which may assist in the decision-making. The real problem is therefore whether automated, 'intelligent' systems exist or not; if they do, situations may always arise where their use is deemed necessary, not because confidence in them is high, but simply because there is no better alternative in response to the particular contingency. On any scale of values, this kind of use of automated decision-making must be termed 'bad'.

The negative aspects associated with the use of computers for decision-support become apparent to a smaller or larger extent depending on the particular type of information processing performed. A typical example, not involving any distinctive 'intelligent' elements, occurred in connection with the space shuttle accident in January 1986; the quantity of data transmitted from the rocket boosters was such that the information pertaining to the malfunctioning components was not displayed in real time to the controllers, and it could only be retrieved later from data tapes. While the accident probably could not have been averted even if the information had been immediately available to the controllers, the example shows that, in some situations, information may be so abundant that some has to be put aside to avoid overcharging the system.

It is precisely this problem of the 'information explosion' that the use of expert systems is addressing; information is processed and compressed to present the decision-maker with simplified choices. Unfortunately, the risk of bad decisions increases when there is a lot of raw data, when decision times are short and when the computer software is inadequate. The risk is particularly high in the complex real-time environment where most of the semi- or fully automated military command and control systems, which are currently being considered by military planners, are supposed to operate. Bad decisions could, for example, result from misinterpretations of the threat posture of an adversary and, as a result, pre-emptive action out of proportion to the actual threat could be initiated. In the extreme case, consequences could be dramatic if authority to release nuclear weapons were involved.

In strategic defence systems, the 'no-decision-time' problem is particularly acute. The solution to the complicated command, control and battle management of wide-scale defensive technologies has therefore repeatedly been argued to require sophisticated computing capabilities and AI techniques. The various inadequacies of these, in turn, raise many concerns about the basic utility of strategic defence and may eventually decide in favour of much less ambitious defensive systems, if any at all, than the ABM systems allowed today.

In summary, the military applications of advanced computing and AI techniques, which in some quarters are envisaged to have a profound effect on command and control structures, tactical or strategic, and to change radically the conduct of war, must be looked upon with both scepticism and concern. The technical feasibility of large-scale, complex projects is very uncertain and the spectre of automation in decision-making leading to crisis instability, accidental war and run-away escalation cannot be discarded. In questions touching international security and armed conflict, decision-makers are likely to be much more comfortable with human intelligence, however inadequate it may be in many situations, than allowing 'intelligent' machines to make fatal decisions. The logical conclusion of this preference is clear: it is unwise to put into force weapon systems and doctrines which in some, more or less improbable, situations of tension and crisis would imply such short decision times that resort to automated systems, using computers and AI techniques, appears to be the only option.

On the positive side of the balance sheet, there are a number of applications involving computers and AI techniques which have a reasonable chance of being effective and sometimes even useful. They are primarily characterized by a certain robustness due to a limited domain of application and/or by decision times which are long enough to permit humans to have an interactive role. In the military area applications in data-base management, technical maintenance and surveillance and some battlefield robotics might be of this kind. For example, robots designed to operate in minefields might become quite effective; such military applications are, however, very far from signalling any major changes in either tactics or strategy.

In arms control applications involving verification, negotiation, modelling and simulation, decision times are relatively long and there is good reason to believe that AI techniques could become an important decision-making instrument in these areas. A common feature of such international-security related applications is the vast quantity of information which must be treated using both quantitative and qualitative assessment; substantial compression and processing of this information is needed to speed up data analysis and decision-making. In a number of important arms control areas, the computer, in combination with expert-system and AI techniques, is therefore likely to have an interesting potential for becoming an essential tool with many user-friendly, 'intelligent' features. The fact that, in this kind of application, humans will always be directly involved in the decision-loop, gives reason to believe that the information processing involved will become both faster and more reliable.

Notes and references

[1] Feigenbaum, E. A. and McCorduch, P., *The Fifth Generation* (Signet: New York, 1984).

[2] Zemanek, H., 'The human being and the automaton', in Mumford, E. and Sackman, H. (eds), *Human Choice and Computers* (North-Holland Publishing Company: Amsterdam, 1975), p. 3.

[3] Pylyshin, Z. W., *Computation and Cognition* (MIT Press: Cambridge, MA, 1986).

[4] Hofstadter, D. R., *Gödel, Escher, Bach: An Eternal Golden Braid* (Penguin Books: Harmondsworth, 1979).

[5] Zadeh, L. A., 'Making computers think like people', *IEEE Spectrum*, Aug. 1984, p. 26.

[6] 'Information processing—enhancing the thought process', *Miltronics*, Feb. 1986, p. 25.

[7] Peckham, J., 'When machines have ears', *New Scientist*, 4 Dec. 1986, p. 54.

[8] Hodges, A., *Alan Turing: The Enigma of Intelligence* (Unwin Paperbacks: London, 1983).

[9] Turing, A. M., 'Computing machinery and intelligence', *Mind*, vol. 59, no. 236 (1950).

[10] Luce, R. D. and Raiffa, H., *Games and Decisions* (Wiley: New York, 1957); Raiffa, H., *The Art and Science of Negotiation* (Harvard University Press: Cambridge, MA, 1982); Brauch, H. G. and Clarke, D. L. (eds), *Decisionmaking for Arms Limitation* (Ballinger: Cambridge, MA, 1983); Axelrod, R., *The Evolution of Cooperation* (Basic Books: New York, 1984).

[11] Chacko, G. K., *Computer-Aided Decision-Making* (American Elsevier Publishing Company: New York, 1972).

[12] Hinckley, R. H., 'National security in the information age', *Washington Quarterly*, vol. 9, no. 2 (Spring 1986), p. 125.

[13] Feigenbaum and McCorduch (note 1).

[14] Labadie, J. R., 'Industrial computer dependence', *National Defense*, Sep. 1986, p. 28.

[15] Lamb, J., 'US and Britain tangle over supercomputers', *New Scientist*, 29 May 1986, p. 18.

[16] 'Supercomputing for the masses', *Defense Electronics*, Apr. 1986, p. 34.

[17] D'Ambrosio, B., 'Expert systems: myth or reality', *Byte*, Jan. 1985, p. 275; Schindler, M., 'Expert systems', *Electronic design*, 10 Jan. 1985, p. 114.

[18] Lemmon, H., 'Comax: An expert system for cotton crop management', *Science*, 4 July 1986, p. 29.

[19] Hannon, J., 'Military computing', *International Defense Review*, Sep. 1985, p. 1439; see also 'Military computing—movement in the market', *International Defense Review*, Sep. 1985, p. 1359.

[20] Din, A. M., 'Strategic computing', in SIPRI, *World Armaments and Disarmament: SIPRI Yearbook 1986* (Oxford University Press: Oxford, 1986), p. 181.

[21] Gerenscher, M. and Smetek, R., 'Artificial intelligence on the battlefield', *Military Technology*, June 1984, p. 86; Dugdale, D., 'Computers in combat', *Defense Electronics*, June 1986, p. 99.

[22] Meyrowitz, A. L., 'Military applications of artificial intelligence', *Signal*, June 1984, p. 45; McCune, B. P., 'The future of artificial intelligence in defense', *Defense Electronics*, July 1985, p. 114; Shumaker, R. P. and Franklin, J., 'Artificial intelligence in military applications', *Science*, June 1986, p. 29.

[23] See Din (note 20).

[24] Bussert, J., 'Deploying very large-scale integrated circuits', *Defense Electronics*, Sep. 1985, p. 159; Castellano, R. N., 'VHSIC program spurs US IC technology', *Defense Electronics*, July 1986, p. 114.

[25] Podell, A., 'GaAs applications for the future', *Defense Science and Electronics*, Oct. 1986, p. 54; Brody, H., 'Ultrafast chips at the gate', *High Technology*, Mar. 1986, p. 28; Anning, N. and Hebditch, D., 'New chip displays its powers', *New Scientist*, 20 Mar. 1986, p. 43.

[26] Rudie, N. J., 'Transient radiation effects in electronics', *Defense Science & Electronics*, Jan. 1986, p. 77; Hall, R., 'CMOS design yields radiation hardness', *Defense Electronics*, June 1986, p. 197.

[27] Neff, J. A., 'Optical computing at DARPA', *Defense Science & Electronics*, Feb. 1986, p. 43.
[28] Almasi, G. and Harvey, S., 'An introduction to parallel processing', *Journal of Electronic Defense*, May 1986, p. 31; see also 'Parallel computer unveiled', *Nature*, vol. 321 (8 May 1986), p. 103.
[29] Geysenheymer, S. J., 'The greatest challenge: man-machine interfaces', *Military Technology*, Oct. 1985, p. 16.
[30] Dinitto, S. A., 'Software engineering: problems and progress', *Journal of Electronic Defense*, Aug. 1986, p. 41; Wertheim, M., 'Software management', *Journal of Electronic Defense*, Aug. 1986, p. 33; 'STARS, Software Technology for Adaptable Reliable Systems', *ACM Sigsoft Engineering Notes*, no. 2 (1983).
[31] Canaan, J. W., 'The software crisis', *Air Force Magazine*, May 1986, p. 46; Newport, J. P., 'A growing gap in software', *Fortune*, 28 Apr. 1986, p. 132; 'Taking a hard line on software', *Defense Electronics*, July 1986, p. 74.
[32] Barnes, J. G. P., 'Ada—The standard language for software engineering', *Military Technology*, May 1986, p. 68.
[33] Fisher, J. E., 'Is the Ada bandwagon rolling at last?', *Signal*, Oct. 1986, p. 57.
[34] Naedel, D., 'Ada and embedded AI', *Defense Electronics*, Apr. 1986, p. 90.
[35] Nilsson, N. J., *Principles of Artificial Intelligence* (Tioga: Palo Alto, CA, 1980).
[36] Andriole, S. J. (ed.), *Applications in Artificial Intelligence* (Petrocelli Books: Princeton, NJ, 1985).
[37] Davis, D. B., 'Artificial intelligence enters the mainstream', *High Technology*, July 1986, p. 16.
[38] Hayes-Roth, F., Waterman, D. A. and Lenat, D. B. (eds), *Building Expert Systems* (Addison-Wesley: Reading, MA, 1983).
[39] Crowder, S., 'Exploring expert systems', *Signal*, Sep. 1986, p. 65.
[40] Schultz, J. B., 'TEMPEST products and orders flourish', *Defense Electronics*, Jan. 1986, p. 101.
[41] Schultz, J. B., 'Ruggedized computers offer low-cost readiness', *Defense Electronics*, Jan. 1986, p. 69; Dugdale, D., 'Computers in combat', *Defense Electronics*, June 1986, p. 99.
[42] Greer, T. H., 'Artificial intelligence: A new dimension in EW', *Defense Electronics*, Sep. 1985, p. 190; O'Shannon, D., 'Expert systems in EW', *Defense Science & Electronics*, Mar. 1986, p. 19.
[43] Kennedy, T., 'Advances in smart munitions', *Defense Science & Electronics*, Oct. 1986, p. 63; Williams, P., 'The role of smart munitions—the impact of autonomous precision munitions', *NATO's Sixteen Nations*, Nov. 1986, p. 50; Wilke, K. H., 'Microwave sensors for intelligent munitions', *Military Technology*, May 1986, p. 32.
[44] 'NASA details expert system needs', *Advanced Military Computing*, 10 Mar. 1986, p. 4.
[45] Rhea, J., 'DARPA pushes strategic computing technology', *Defense Electronics*, June 1984, p. 113; Klass, P. J., 'DARPA envisions new generation of machine intelligence', *Aviation Week & Space Technology*, 22 Apr. 1985, p. 46; Judge, J. F., 'SCP gets high marks at midterm', *Defense Electronics*, May 1986, p. 65; Ornstein, S. M., Smith, B. C. and Suchman, L. A., 'Strategic computing', *Bulletin of the Atomic Scientist*, Dec. 1984, p. 11.
[46] Shaker, S. M. and Wise, A. R., 'Walking to war', *National Defense*, Mar. 1986, p. 59.
[47] Fuglsang, E. J., 'Robots on the battlefield', *Defense Electronics*, Oct. 1985, p. 77; Gerencser, M. and Smetek, R., 'Robotics and robotic vehicles—the state of the art', *Military Technology*, May 1986, p. 74; Welling, W., 'Battlefield robotics: The beginings of a new defence initiative', *Military Technology*, Oct. 1986, p. 80.
[48] Morishige, R. I. and Retelle, J., 'Air combat and artificial intelligence', *Air Force Magazine*, Oct. 1985, p. 91; Broadbent, S., 'Changing the face of the modern cockpit', *Jane's Defence Weekly*, 1 Mar. 1986, p. 389; Retelle, J. P. and Kaul, M., 'The pilot's associate—Aerospace applications of artificial intelligence', *Signal*, June 1986, p. 100.
[49] Crutchley, M. J. and Lynn, D. G., 'Emerging technologies for forward defence', *Military Technology*, Oct. 1986, p. 44.
[50] Zraket, C. A., 'Strategic command, control, communications and intelligence', *Science*, vol. 224 (22 June 1984), p. 1306; Blair, B., 'Strategic command and control and national security', *Signal*, March 1985, p. 23; Fidelman, M. R., Herman, J. G. and Baum, M. S., 'Survivability of the defence data network', *Signal*, May 1986, p. 148.
[51] 'Automating the Joint Chiefs of Staff', *Defense Science & Electronics*, Feb. 1986, p. 27.
[52] Pengelley, R., 'HEROS and Wavell—battlefield ADP enters a new era', *International*

Defense Review, Oct. 1986, p. 1459; Otis, G. K. and Driscoll, R. F., 'Making the C^3 pieces fit in Central Europe', *Signal*, Nov. 1986, p. 19.

[53] Boutacoff, D. A., 'Army banks on joint STARS for AirLand Battlemanagement', *Defense Electronics*, Aug. 1986, p. 77.

[54] Leonardis, S. and Semprini, M., 'Programming environments for C^3 system development', *Signal*, Oct. 1986, p. 29; Lenat, D. B. and Clarkson, A., 'Artificial intelligence and C^3I', *Signal*, June 1986, p. 115; 'AI in C^3, a case in point', *Signal*, Aug. 1986, p. 79; Bentz, B., 'An automatic programming system for signal processing applications', *Pattern Recognition*, vol. 18, no. 6, 1985, p. 491.

[55] Powell, G. M., Loberg, G., Black, H. H. and Gronberg, M. L., 'ARES: Artificial intelligence research project', *Signal*, June 1986, p. 106.

[56] Gaertner, W. W., '$(AI)^2$ for real time', *Signal*, Apr. 1985, p. 55; Drogin, E. M., 'AI issues for real-time systems', *Defense Electronics*, June 1986, p. 150.

[57] Chapman, G., 'AirLand Battle doctrine and the Strategic Computing Initiative', *The CPSR Newsletter*, vol. 3, no. 4 (Fall 1985).

[58] Chodakewitz, S. B., Deane, M. J. and Weatherly, S. G., 'Soviet strategic C^3: The threat and counterthreat', *Signal*, Dec. 1985, p. 79; 'Soviet C^3I', *Defense Science & Electronics*, Sep. 1986, p. 45.

[59] Southern, J. R., Davis, C. G. and Edwards, M. P., 'Army BM/C^3 in the SDI program', *Signal*, July 1985, p. 37; Whelan, C. R., 'Developing the technology for the SDI battle management/C^3 system', *Signal*, July 1985, p. 44; Lacer, D. A., 'C^3 for strategic defense', *Journal of Electronic Defense*, July 1985, p. 37; Offutt, J. H., 'SDI system architecture and battle management/C^3 program', *Journal of Electronic Defense*, July 1985, p. 45; Grant, D., 'Battle management for SDI—C^3 on a global scale', *Military Technology*, Oct. 1986, p. 163.

[60] Abrahamson, J. A., 'Electronics and the strategic defense initiative', *Journal of Electronic Defense*, July 1985, p. 27; Rankine, R. R., 'SDI and role of microcircuits', *Defense Science & Electronics*, Jan. 1986, p. 21.

[61] Blau, T., Goure, D. and Hopkins, K., 'Simulation and SDI', *Military Technology*, Jan. 1986, p. 4.

[62] Lin, H., 'Software for ballistic missile defense' (MIT: Cambridge, MA, June 1985); Nelson, G. and Redell, D., 'The star wars computer system' (Computer Professionals for Social Responsibility: Palo Alto, CA, June 1985); Parnas, D. L., 'Why the SDI system will be untrustworthy' (University of Victoria: Vancouver, June 1985); Parnas, D. L., 'Artificial intelligence and the strategic defense initiative' (University of Victoria: Vancouver, June 1985).

[63] Eastport Study Group on Strategic Computing, Report to the Strategic Defense Initiative Organization, Dec. 1985.

[64] Tsipis, K., Hafemeister, D. W. and Janeway, P. (eds), *Arms Control Verification* (Pergamon-Brassey's: Washington, DC, 1986); Krass, A., SIPRI, *Verification: How Much Is Enough?* (Taylor & Francis: London, 1985).

[65] Liu, H. H., 'A rule-based system for automatic seismic discrimination', *Pattern Recognition*, vol. 18, no. 6 (1985), p. 459.

[66] Din, A. M. and Sauder, A., 'The prospects for a satellite monitoring agency', in Avenhaus, R., Huber, R. K. and Ketelle, J. D. (eds), *Modelling and Analysis in Arms Control* (Springer: Berlin, 1986), p. 391; Jasani, B. and Sakata, T. (eds), SIPRI, *Satellites For Arms Control and Crisis Monitoring* (Oxford University Press: Oxford, 1987).

[67] 'DMA stresses LANDSAT, AI, real-time ops', *Military Space*, 29 Apr. 1985.

[68] Avenhaus, R., Huber, R. H. and Ketelle, J. D., 'Systems analysis and mathematical modelling in arms control', *OR Spektrum*, vol. 8 (1986), p. 129.

[69] Davis, P., 'Rand's experience in applying artificial intelligence techniques to strategic-level military-political war gaming', Rand Corporation Report, Santa Monica, Apr. 1984.

[70] Ward, M. D., 'Simulating the arms race', *Byte*, Oct. 1985, p. 213.

[71] Brams, S. J., *Superpower Games* (Yale University Press: New Haven, CT, 1985); Fraser, N., 'General ordinal 2×2 games in arms control applications', in Avenhaus, Huber and Ketelle (note 66), p. 307; Fichtner, J., 'On concepts for solving two-person games which model the verification problem in arms control', in Avenhaus, Huber and Ketelle (note 66), p. 421.

[72] Bonhan, G. H., 'Simulating international disarmament negotiations', *Journal of Conflict Resolution*, vol. 15, no. 3 (1971), p. 299.

Part II. Artificial intelligence concepts and computer technology

Chapter 2. An introduction to artificial intelligence[1]

ROBERT DALE

I. Introduction

Although computers have only been with us for around 40 years, the pace of technological development during this period has been rapid. This is most obvious to the non-specialist simply in terms of the size of computer hardware: the first computers were large enough to fill entire rooms, but today, microprocessor technology has put computers not only on people's desks but also on their laps. Now even wristwatches and electric toasters contain microchips and proudly proclaim 'intelligence'.

Less visible than these advances in hardware, but just as important, are the advances that have been made in computer software. As the hardware has become smaller, it has also become cheaper and faster. This has allowed more complex tasks to be tackled, and this in turn has required correspondingly more complex programs to be written. As a result, considerable developments in software techniques have taken place, although these have always tended to lag behind the advances in hardware.

One area of advanced computing which has attracted a great deal of interest in recent years is *artificial intelligence* (AI).[2] Put briefly, the aim of AI researchers is to construct programs which enable machines to behave 'intelligently'. The aim of this paper is to give some understanding of what this means. We begin by sketching a definition of AI, followed by an explanation of what AI is in terms of the tools, techniques and applications it deals with. We then discuss a particular application of AI, *expert systems*, in some detail. Finally, we make some suggestions as to what the future holds for AI.

II. What is AI?

Definitions of AI

Artificial intelligence is very difficult to define satisfactorily. This is partly because it is many things to many people, and partly because its subject matter, intelligence, is similarly difficult to define. Unfortunately, the difficulty of providing a concise definition has only served to maintain the mystique that sometimes surrounds the subject.

Although a simplification, it is helpful to partition researchers in AI into two different groups. The first of these, the cognitive scientists, are fundamentally interested in understanding how the human mind works. Their principal assumption is that the mind can be conceived of as an information-processing mechanism, and so the computer serves as a good model for testing their theories; their goal is to build an intelligent machine, and in so doing, to understand what intelligence is.

The second group of AI researchers are less concerned with psychology, and more concerned with engineering. Their goal is the development of computer programs that behave in ways that are 'smarter' than conventional software. Such systems are described as 'intelligent' by their makers, and others in the field, insofar as they carry out tasks which, if they were performed by people, we would want to say require intelligence. Of course, this is not the same as saying that the full breadth of human intelligence is being emulated: such programs are 'intelligent' only with respect to very narrowly specified tasks.

For this second group of researchers, there is no requirement that the programs actually perform the tasks in question by using the same techniques as are used by people (although, of course, the study of human intelligence can provide ideas as to how such goals might be achieved); the primary goal is that the *external behaviour* is in some way 'intelligent', irrespective of how that behaviour is achieved.

For our purposes, it is the second approach to AI that is important. Regardless of whether or not AI leads to a better understanding of the human mind, it does not seem clear that it will lead to a new generation of advanced computer systems. It cannot be emphasized strongly enough, however, that AI programming is not immune from the problems that beset more conventional programming. In particular, the same rule of 'garbage in, garbage out' applies: a program can only do what it has been instructed to do by its programmer, and if the programmer's conception of the task is in error, then the program is unlikely to function as required. AI programs do not possess any 'emergent intelligence'. In fact, the situation in AI is much worse than in conventional programming, for the simple reason that AI programs attempt to model many processes, such as language use and reasoning ability, which no one really understands.

We can get a better idea of what AI involves by looking at the tools and techniques that are used in AI, and the applications they are used to construct.

AI tools

Although computers ultimately only 'understand' strings of ones and zeros, it is extremely rarely that a programmer will communicate with a computer in such terms. The task of programming computers has been made easier by the development of a large number of high-level languages; two widely known examples are Basic and Pascal. These languages are much easier for humans to use and understand, but programs written in them can be subsequently translated into code suitable for execution on the machine.

In principle, all programming languages are equivalent in the sense that a program written in one language could be translated into a program written in any other. However, each particular programming language is usually designed with a certain purpose in mind, making it easy to express particular sorts of operation more efficiently.

It is often said that one of the fundamental differences between AI and conventional programming is that, whereas the latter is based on numeric calculation, the former is based on the notion of *symbol processing*. It is not surprising, then, that special symbol-processing languages have been developed for AI programming. The two most well known of these are Lisp, which came into being in the 1950s, and Prolog, a more recent development.

Surrounding these languages, a significant amount of research in AI is oriented towards the development of programming *environments*, or integrated sets of tools, that ease the programmer's task. These environments help overcome at least some of the problems that arise in the development of the extremely complex programs to be found in serious AI applications. Specialized hardware is also available, utilizing particular computer architectures which can process AI languages more efficiently.

An important aspect of AI is the particular style of programming that has developed, sometimes referred to as *exploratory programming*. Unlike the development cycle of conventional software, where a detailed design specification will be written before any implementation, much AI work is exploratory in nature. The programmer constructs a program without knowing whether it will perform the required task. In doing so, he or she typically discovers some unforeseen problems, and amends the program accordingly; this process repeats in a cycle of continual refinement. This method of programming is not conducive to well-structured or easily maintainable programs, and in theory any program written in such a fashion should be subsequently rewritten in a more structured manner; however, this typically does not happen.

AI techniques

At the core of almost all the sub-fields of AI are the techniques of search and knowledge representation. These concepts are themselves the focus of a great deal of research, and so we can only hope to scratch the surface in this exposition.

Search

In the very early days of AI, the attempt to construct computer programs which could play games such as chess led to the realization that many such problems could be characterized as the problem of searching for a solution within a set of possibilities, known as a *state space*.

In this approach, a problem is characterized in terms of an *initial state*, a desired *goal state*, and a set of *operators* that perform the transition from one state to another; the problem is solvable if the goal state can be reached from the initial state by repeated applications of the available moves in an appropriate order.

The state space can be viewed as a network in which the states are nodes, and the operators are the arcs between the nodes. This is most easily explained with respect to a simple game such as 'noughts-and-crosses'. Figure 2.1 shows the initial state of play, and the different states that can be reached from this state in one move. From each of these possible second states, another set of possible third states can be generated, and so on; the figure shows the expansion of only a single move at each of the first three stages, since there is not enough space on the page to show the complete expansion even for the *first* move.

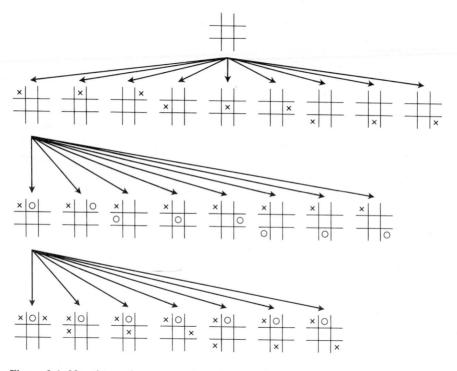

Figure 2.1. Noughts-and-crosses: exhaustive search

Obviously, the size of the network increases dramatically with each successive level considered, a phenomenon referred to as the combinatorial explosion. In the case of noughts-and-crosses, there are 9!—i.e., 362 880—possible states in the state space.

Initial attempts to solve problems characterized in this way involved the brute force method of *exhaustive search*; a number of strategies were developed for considering each node in the network until the goal state was found. For many problems, however, this approach is hopeless, because of the large number of states to be considered: in chess, for example, there are 10^{120} possible paths in the state space.

Awareness of these problems led to the development of more sophisticated methods known as *heuristic search*: instead of searching the state space blindly, various rules of thumb are used to constrain the search to those parts of the state space that, on the basis of some *evaluation function*, look most promising. One such evaluation function for our simple noughts-and-crosses example would be to select the position through which there are the most winning paths, as is shown in figure 2.2 for the first move of each player.

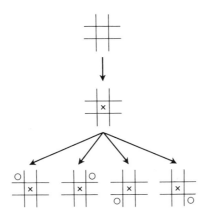

Figure 2.2. Noughts-and-crosses: heuristic search

The use of heuristics rather than algorithms is another defining characteristic of AI. Conventional programming involves the design of algorithms which are guaranteed to give the correct solution in all applicable circumstances; heuristics, on the one hand, are only correct most of the time.

Knowledge representation

The importance of work in knowledge representation was also realized very early in AI's history. Just as characterizing a problem as a case of state-space search provides a way of applying problem-solving techniques, the choice of an appropriate way of representing knowledge for a problem can vastly simplify the solution of that problem. Every representation emphasizes certain information about a concept and ignores other information: choosing a good representation makes the right information available for the problem's solution. In our noughts-and-crosses example, we could have reduced the number of states to be considered by using a representation that allowed us to make use of the symmetry in different states: as shown in figure 2.3, this further reduces the number of best moves at the second round of the game from four to one.

Semantic networks are probably the most widely known knowledge

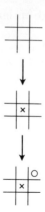

Figure 2.3. Noughts-and-crosses: simplified solution

representation formalism. In this notation, knowledge is represented as nodes connected by arcs, both of which can be labelled: the nodes represent objects, concepts or situations in the problem domain, and the arcs represent the relationships between the nodes. Thus, as in figure 2.4, we can represent the three facts that (*a*) Clyde is an elephant, (*b*) elephants are grey in colour, and (*c*) elephants are mammals. This representation makes it very easy to deduce facts that are not explicitly stated in the network: in our example, we can easily deduce that Clyde is grey and that Clyde is a mammal.

Another frequently used knowledge representation formalism in AI is the

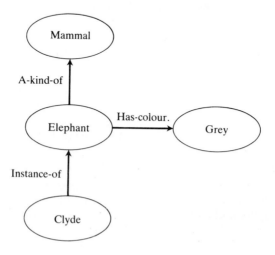

Figure 2.4. Semantic networks

INTRODUCTION TO ARTIFICIAL INTELLIGENCE

frame-based system. This is similar to the basic semantic network formalism, except that each node in the network now has internal structure: it is a frame consisting of a number of attributes or slots (e.g., *colour, size, material*), each of which contains a value (e.g. *large, red, wood*). Each slot can also have arbitrary pieces of computer code attached, to be executed under specified circumstances: for example, when information in the slot is added, deleted or requested. So, for example, when a request is made for information in a particular slot, we may have a piece of code which effects some changes elsewhere. The frame approach lends itself neatly to the development of mechanisms where nodes in the network inherit default values for slots by virtue of the values held in superordinate frames. Figure 2.5 shows a network of frames which describe some items of furniture. We have three nodes: *table* represents the *concept* of a table, and provides the defining characteristics of tables (very much simplified of course); *table-1* and *table-2* are two particular *instances* of the concept. Notice that *table-1* inherits its *number of legs* from the superordinate node by default, whereas the default value is overridden for *table-2*; and *table-2* inherits its *material* from the superordinate node, whereas *table-1* overrides this default value.

Most work in knowledge representation involves some elaboration of these relatively simple ideas. However, our simple examples avoid the complexity of the knowledge representation task for any realistic problems. The task of

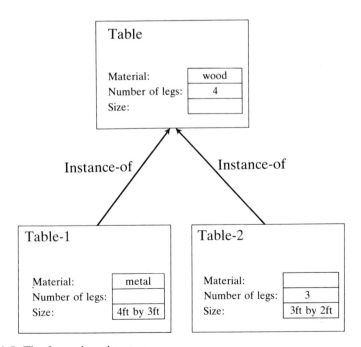

Figure 2.5. The frame-based system

constructing an appropriate knowledge representation is a very difficult one, and the best solution typically depends on the particular problem to be solved. Analogously to the problem of search, it is often the case that in order to solve a specific problem, a form of knowledge representation which is specific to the problem domain may be required.

AI applications

By applications, we mean here the particular tasks addressed by AI programs. There are a number of particular application areas that are conventionally seen as being within the domain of AI, the principal ones being: problem solving, logical reasoning, learning, natural-language processing, robotics and vision.

Work in problem solving goes way back to the beginnings of AI where, as we saw earlier, much interest focused upon the issues involved in constructing programs that could solve puzzles or play games such as chess. Today, there are programs which can perform to very high standards in these areas, and these are being constantly improved. Research in this area is closely related to work in planning, where AI techniques are used to develop plans that achieve desired goals.

Work in logical reasoning is closely related to the work in problem solving. This area focuses on the construction of systems that reason and draw inferences from a data base of facts. This work has resumed some of its earlier prominence because of the resurgence of interest in formal correctness and reliability. Related work is concerned with 'truth-maintenance systems', and the issues that arise when there are changes in the truth or falsity of the facts on which inferences have been based.

Learning is also an active area of interest in AI, but it is one which has not had a great deal of success. Some expert systems 'learn' rules from examples, although the processes involved here are far from what we would want to call learning in human beings. Learning algorithms are based on pattern matching and inductive generalization, that is, the formulation of rules by abstracting out similarities and differences between instances. Such programs still have to be told exactly which dimensions of a situation are to be considered, and so are not particularly sophisticated.

Natural-language processing has been a major area of AI research from early on. There are a number of sub-areas in this field: most work draws upon research in linguistics and has focused upon the issues involved in the understanding of written language, although more recently there has been increased interest in the production of language. Current systems can 'understand' simple sentences in specific, predefined contexts, where by 'understand' we mean that the system responds to a manner which is deemed appropriate. There is also related work in speech recognition, which relies more heavily on research in signal processing. State-of-the-art systems in this area can typically deal with either a considerable vocabulary when trained for specific speakers, or a small vocabulary for a larger number of speakers, but not both.

Work in robotics has two distinct aspects. One part of this research is very engineering-oriented, and is concerned with the processing involved in actually manipulating physical devices in very precise ways so that, for example, not too much pressure is applied to objects when they are grasped. The other aspect of this work focuses on the construction of plans of action, and is part of the work on planning mentioned above. Existing systems can carry out relatively simple assembly tasks.

As in the case of speech recognition, much work in vision concerns signal-processing issues. Principal concerns here are tasks such as getting a machine to recognize shapes and objects. So far, recognition of simple shapes is possible, but features of real-life situations such as shadows cause considerable problems.

The state of the art in almost all sub-fields of AI is such that truly practical results are a long way off. So far, the only sub-field that has achieved anything approaching widespread recognion is expert-systems technology; and so we describe this sub-field in greater detail in the next section.

III. Expert systems

What is an expert system?

Expert systems[3] are programs which represent and apply knowledge within some area of expertise, with the purpose of solving problems or giving advice in that domain. They are referred to as being expert because they embody levels of experience equivalent to or greater than that of human experts in those areas. In practice, the term 'expert system' is also used to describe what should more properly be called *knowledge-based systems*: like expert systems, these are programs that embody explicitly encoded knowledge, but their capabilities are more restricted, and do not merit the appellation 'expert'.

The basic idea underlying expert systems is that to make a program 'intelligent', you must provide it with lots of high-quality, specific knowledge about some problem area. This approach is merely the latest in a series of changes of emphasis within AI. In the early days of AI, the major emphasis of research was on the development of general techniques that could be used in a wide range of problems. However, it was eventually realized that general principles could not easily be abstracted out from the particular problems considered, and the results which were obtained were difficult to scale up to more realistic problems. Thus, the years from 1970 onwards saw a re-orientation within the AI-research community. The focus of research shifted to the development of general techniques that could be used to create specialized systems. The problems considered were now much narrower, and more use was made of context-dependent information about those problems. The development of expert systems is the next natural step in this progression, with the explicit acknowledgement of the utility of separating out the program's knowledge from the way that knowledge is used.

The basic architecture of an expert system

The principle components of an expert system are an *inference engine* and a *knowledge base*. The inference engine is a control mechanism that uses the information in the knowledge base to solve the problem under consideration. It may use its knowledge of the situation to suggest a plan of action for the user; in an expert system which acts merely as an assistant or an advisor, the user can decide whether or not to do as the program suggests. In certain time-critical applications, however, the human may be 'removed from the loop', and the system allowed to act autonomously.

The knowledge base contains two kinds of information: global information about the domain and specific information about the particular problem under consideration. The inference mechanism determines how to use this general domain knowledge in conjunction with the specific problem information to produce a solution.

Building an expert system

The general process of constructing an expert system is referred to as *knowledge engineering*. The task of constructing the inference mechanism requires that the designers of the system decide which control strategies are most suitable for the application of rules in the domain of application. The process of building the knowledge base is often seen as the most difficult part of constructing an expert system. The knowledge engineer consults with an expert in the application area, with the aim of extracting his or her expertise: the procedures, rules of thumb and factual knowledge that the expert uses. The fundamental problem here is that, typically, experts do not think of their knowledge in the explicit manner necessary in order to encode it in a computer system. The knowledge engineer will therefore interview the expert and observe his or her behaviour in carrying out the appropriate task in order to determine what procedures are being used. After a considerable amount of time has been invested in this exercise, the knowledge engineer can then encode the results in a working system. The task is not then complete, however; typically, as it is used, the expert system will be found to be inadequate, in that its behaviour and predictions do not always match those of the expert. This leads to a cycle of continual refinement, as the knowledge base of the system is updated to improve its performance.

How an expert system works

To see how an expert system works, we can consider the most popular approach to their construction. In this aproach, knowledge is represented as 'if-then' rules. The general form of these is such that, if a number of conditions hold, then some other condition can also be said to hold, or some action can be taken on the basis of these conditions holding. A simple example is

IF (the power supply on the space shuttle fails) AND (a back-up power supply is available) AND (the reason for the first failure no longer exists) THEN (switch to the back-up power supply)

In a rule-based expert system, the inference mechanism decides which rules to apply on the basis of the current situation; the results of applying the rules are noted, and other rules are applied as a result. The sequence of application of the rules then amounts to a chain of reasoning. Thus, given some initial evidence, the program can work towards a solution: this approach is referred to as *forward inference*. Alternatively, given some hypothesized solution or desired goal, the program can apply the rules backwards to see if there is evidence to support the hypothesis: this is known as *backward inference*. Each approach is suitable for different types of problem; some systems mix both approaches.

Some examples of expert systems

Research into expert systems has been carried out in a large number of areas, including agriculture, chemistry, computer systems, electronics, engineering, geology, information management, law, manufacturing, mathematics, medicine, meteorology, military science, physics, process control and space technology. However, very few expert systems have reached a mature stage of development where they are in full operational use. When expert systems are discussed in the literature, the same small set of half-a-dozen or so systems are always used as examples. The early show-piece systems usually discussed are DENDRAL, MYCIN and PROSPECTOR.

DENDRAL,[4] on which work began in 1965, determines the molecular structure of organic compounds using mass-spectral and nuclear-magnetic response data. It systematically enumerates all possible molecular structures consistent with the data, and then uses its knowledge of chemistry to cut down the list of possibilities to a more manageable size. It functions primarily as an assistant to the user, allowing him or her to make decisions when the program does not know what to do.

MYCIN,[5] whose development began in 1972, performs medical diagnosis. It contains knowledge relating infecting organisms in the blood with case histories of patients, known symptoms and laboratory test results, on the basis of which it can recommend appropriate treatment.

PROSPECTOR,[6] begun in 1974, aids exploration geologists in searching for ore deposits. The geologist inputs to the program details of an area of interest. The program compares this information with its internally held models of existing ore deposits; this allows the program to recognize similarities and differences, and to note what other information it needs. It then carries out an information-seeking dialogue with the geologist to obtain this additional information, on the basis of which it can then predict the mineral potential of the area in question.

More recently, the first commercially successful expert system was the Digital Equipment Corporation R1,[7] on which work began in the late 1970s. This system takes a customer's VAX minicomputer order and checks it to see if it contains all the necessary components. It then determines the spatial arrangement of these components, which it outputs as a diagram for use by the technicians who assemble the systems. R1 is widely recognized as the largest and most developed rule-based expert system in operation.

In the area of military applications, the use of expert-systems technology has been considered in the following areas: aircraft identification, automation of emergency procedures in fighter aircraft, radio/radar targeting aids, processing of sensor data, weapons targeting, mission planning, aircraft maintenance, ocean surveillance, and target recognition. The US Department of Defense's Strategic Computing Program (SCP), launched in 1983, consists of a 10-year period of research with the aim of utilizing advanced computing for military applications, with $600 million allocated for the first five years. The original plan states the need for the development of an appropriate technology base—towards the development of systems which exhibit 'human-like, intelligent capabilities'—and defines three particular applications to be pursued: an autonomous land vehicle (for the US Army); a pilot's associate (for the Air Force); and a battle-management system (for the Navy). All of these are expected to rely to some extent on expert-system techniques.

Problems of expert systems

Although expert-system technology may appear to offer great promise, it is important to realize that the current state of the art is not as well developed as some enthusiastic proponents might suggest.

There are, for example, substantial unsolved problems in the construction of expert systems. There are all the problems associated with acquiring and encoding the knowledge for a given expert system in the first place, a task often seen as the bottleneck in expert-system construction. Also, no real methodology for the construction of expert systems has been developed: there are no guaranteed means of selecting the correct knowledge representation or type of inference for particular tasks. The development time even for relatively simple expert systems is still long; and the effectiveness of operational systems often turns out to be much less than when these systems are tested in laboratory conditions.

We should also be aware of the limitations of existing expert-systems technology. Expert systems are completely uninspired, and have no creativity; they do not know how to handle events which fall outside the class of events foreseen by their designers; they only operate in very narrow domains of expertise; they exhibit very fragile behaviour at the boundaries of their knowledge; and they possess very limited explanation capabilities. All these are quite the opposite of the characteristics possessed by human experts, however—a reminder that we should not take the term 'expert system' too literally.

No doubt some of these problems will be overcome in time. However, one of the major lessons to be learned from the development of R1, mentioned above, is that the process of incrementally refining the knowledge base for a full-scale expert system may *never* be complete. This implies that there may *always* be situations in which the system provides the wrong answer; and this is an important consideration to bear in mind when designing systems which have to operate in such a short time-frame that human involvement is ruled out.

IV. Future prospects

What does the future hold for AI? Inevitably there will be hardware improvements: processors will become faster, and memory will become cheaper, resulting in a computational infrastructure that can support larger and more complex programs. In addition, there is increasing interest in *parallel processing*, resulting in computers that can perform a number of tasks simultaneously. One result of this trend will be the development of new algorithms: the replacement of conventional hardware architectures will result in many algorithms which were previously just too computationally demanding now being seen as feasible. We may even see a return to the more brute force methods of computation which characterized some of the very early work in AI. Trends of this sort are likely to lead to considerable advances in application areas such as speech recognition and signal processing.

However, with the increasingly sophisticated hardware come problems. As noted above, advances in computer software have tended to lag behind developments in hardware. Although the hardware supports increasingly complex computations, the methodologies for software development at one level of complexity do not necessarily scale up to the next. Although we now have parallel hardware, we know very little about how it can be used.

There is a more general problem with AI software which is alluded to in our discussion of the limitations of current expert systems. We can draw a useful distinction between *special-purpose* and *general-purpose* systems. All the AI systems that have been built so far are very much special-purpose systems. The successes that have been achieved are based on exactly this property: they operate in very narrowly defined circumstances, not unlike the toy domains typically used in basic AI research. This specificity of application is necessary in order to have the systems operational at all. However, by virtue of not possessing the wide background possessed by humans (usually referred to as common-sense), these special-purpose programs are completely out of their depth as soon as they have to face situations not foreseen by their designers. We do not know how to reliably construct AI systems which have to perform in unpredictable, hostile real-world circumstances.

Unfortunately, there is a real danger of such problems being ignored. The history of AI is littered with the making of grandiose claims, and the resultant problems of raised expectations. This hype, which is not always discouraged by those working in the field, has caused problems in the past: in the mid-1960s, the US DoD cut funding in machine translation work because of the lack of results, and funding for AI research in the UK was drastically cut in the early 1970s because the field was considered not to be living up to its promises. It might be suggested that things are different now. Indeed, there have been substantive improvements in the field. There is also a better awareness of the limitations of the technology. However, many AI techniques which are at the forefront of current research are wrongly presented in the commercial press as state-of-the-art.

In attempting to characterize the current state of AI technology, Bundy[8] has drawn an analogy between AI and the building of bridges in civil engineering. The latter technology now allows the reliable construction of massive structures such as the Golden Gate Bridge, but, by comparison, AI is still at the stage of building village footbridges.

Notes and references

[1] The author would like to thank colleagues in the Edinburgh Computing and Responsibility Group and the University of Edinburgh Centre for Cognitive Science for their helpful comments on an earlier version of this paper.

[2] A good basic text on artificial intelligence is Charniak, E. and McDermott, D., *Introduction to Artificial Intelligence* (Addison-Wesley: Reading, MA, 1985).

[3] A good basic text on expert systems is Waterman, D., *A Guide to Expert Systems* (Addison-Wesley: Reading, MA, 1985).

[4] Buchanan, B. G., Sutherland, G. L. and Feigenbaum, E. A., 'Heuristic DENDRAL: a program for generating explanatory hypotheses in organic chemistry', in Meltzer, B. and Michie, D. (eds), *Machine Intelligence*, Vol. 4 (Edinburgh University Press: Edinburgh, 1969).

[5] Shortliffe, E. H., *Computer-based Medical Consultation: MYCIN* (American Elsevier: New York, 1976).

[6] Duda, R. O., Gaschnig, J. G. and Hart, P. E., 'Model design in the PROSPECTOR consultant system for mineral exploration', in Michie, D. (ed.), *Expert Systems in the Microelectronic Age* (Edinburgh University Press: Edinburgh, 1979) pp. 153–67.

[7] McDermott, J., 'R1: an expert in the computer systems domain', in *Proceedings of the First National Conference of the American Association for Artificial Intelligence*, Los Altos, 1980, pp. 269–71.

[8] Bundy, A., 'AI bridges and dreams', Research Report No. 200, Department of Artificial Intelligence, University of Edinburgh.

Chapter 3. Hardware requirements for artificial intelligence

LENNART E. FAHLÉN

I. Introduction

The increased use of computers for artificial intelligence (AI) and other non-'number-crunching' applications during the 1960s and 1970s soon revealed severe shortcomings in available computer architecture for handling programs with the required memory and computational capacity. Also, the development of software for AI needed a more interactive and flexible environment with, among other things, a much higher communication bandwidth between the computer and the programmer.

'Explorative programming' is a term used for the kind of experimental 'test it and see' work through which many AI-researchers seek optimum (or at least adequate) algorithms or data representations for a problem domain which is not so well understood. During the early 1980s the development of architecture and machines specifically for artificial-intelligence work has shown that major improvements are not obtained exclusively through new software techniques and AI concepts. In fact many of the architectural innovations, both software and hardware, are applicable to conventional computing as well, and there is extensive cross-fertilization. Things like bit-mapped screens, windows and mouse devices are now commonplace; today even inexpensive personal computers provide respectable environments for AI development.

A survey such as this can only point out interesting developments rather than provide an exhaustive treatment of any of the subjects. It is however possible to indicate the present state of the art. Because the area of investigation is large and heterogeneous and a detailed description of some of the issues would be very technical and beyond the scope of this presentation, a reference list for further reading is given at the end of this chapter.

II. Symbolic versus numerical computing

By symbolic computing we mean the ability of a program (or machine) to manipulate symbols, where a symbol designates more than just a number (or a variable cell in the memory) as in conventional numerical computing ('number-crunching'). A symbol can represent anything from the name of a person to a relation, intention or plan—it can be used to represent knowledge in an abstract way. A further distinction of symbolic computing is the common use of non-algorithmic methods (heuristics) to solve problems, whereas in numerical computing algorithmic ('step-by-step') solutions dominate. Also, in symbolic representation of knowledge the semantics of the data are explicit, whereas in conventional representation the meaning is embedded in the procedures that make up the program. A typical difference between conventional programs and programs written for symbolic computing is that the latter handle problem domains (which can of course be very small), rather than specific problems within a domain.

With knowledge-oriented computer architecture, hardware and software, several different levels are needed for representation and processing. Manipulation of large symbolic structures consisting of lists, objects, frames and composites are common. Programs for symbolic processing often have unusual internal structures and use novel programming techniques.

Searching

One example of an operation that is important in symbolic computing is searching. Searching is the investigation of a domain of possibilities (a search tree in game playing, for example), to find an analogy to some other problem area. This operation must be carried out in a systematic and efficient way. Searching algorithms are plentiful; two simple ones are:

1. Depth-first algorithms, in which each possibility (or branch on the search tree) is traversed in turn through to the end—once a possibility has been processed we know if it contains the solution or not, and can then proceed to the next one.

2. Breadth-first algorithms, in which a branch is investigated to a certain (perhaps variable) depth: the state of that search is then saved while the next branch is considered. When the last branch has been processed the algorithm returns to the first, continues from the point at which the search was stopped and processes the branches a bit further, and so on.

It is of course possible to mix these two methods of search. Other important operations useful for AI are deduction of facts, matching of patterns and sorting.

III. Programming languages for AI

Three examples of programming languages used in AI are listed below.

Prolog

Prolog is a programming language based on first-order predicate calculus. Prolog programs have both procedural and declarative aspects. Aside from the pure logic (declarative) part, the (procedural) control component should not influence the meaning of the program but is extremely important for efficiency.

A program consists of a collection of predicates, where each predicate is made up of one or more 'clauses'. Each clause is made up of a head part (also called a conclusion part) and, unless it is a primitive fact, a body part (also called a condition part), which is a collection of names for the head parts of other clauses.

By posing a query (a rule or fact to be proven or refuted) to the system, the execution is started at the top of the program and goes to the bottom trying to find an appropriate clause to prove by matching the head of the clauses with the query using unification (see below). When a clause is found the predicate calls in the body are executed from left to right (becoming new queries unless the body is empty, in which case the query just succeeds). If a call fails, Prolog backs up to the most recent untried alternative in the search path and tries that instead. This is called backtracking and is one of the major features of the language. When the system has run out of alternatives, it returns a fail message to the level above; if this is the system level, the original query returns a fail and the program terminates.

The execution model of Prolog is an example of the use of pattern-directed invocation of procedures as opposed to the conventional explicit (imperative) way of calling procedures.

Unification

What gives Prolog its power is the concept of unification between variables. Unification is a term that comes from predicate calculus and represents an algorithm for matching two expressions. This is done by trying to make the expressions equal by finding a substitution (or binding) for as few as possible of the involved variables. If this succeeds the two expressions are said to unify with each other and the resulting binding (instantiation) of the variables is called the unifier. Variables are single assignment: they are bound by the unification and can be unbound by backtracking, but the bindings are never changed within a computation.

Prolog makes use of the 'closed world assumption': if a goal cannot be proved, it is false—all instances of a relation can be found by searching the data base. This limited view of the world can cause considerable trouble and is something that has to be taken into account when writing Prolog programs.

In Prolog there are many 'extra-logical' operations. These increase the power of Prolog as a programming language but decrease its theoretical soundness and often make it more difficult to use Prolog for parallel processing. Some predicates—assert(X), retract(X), call(X), for example—are especially useful for AI or expert system applications.

Lisp

Lisp is derived from a formal mathematical system called lambda calculus and was invented by John McCarthy (then at MIT) in the late 1950s. One of the oldest programming languages in existence, its original formal soundness has been diluted over the years by the addition of a large selection of extensions to make it more convenient and powerful.

Lisp programs are built by combining simple functions into more complex functions; recursion is very common. Lisp has meta-operations, making it possible for functions to contain and to execute other functions. The fundamental data objects in Lisp are the so-called cons-cell and the atom. A cons-cell is simply a pair of pointers (memory addresses) pointing to other objects in the memory or to NIL, the null-pointer. Atoms are primitive objects such as symbols, variable names and smaller numbers. The most important data structure in Lisp is the list. A list is built from interconnected cons-cells, in such a way that the first pointer in the cons-pair (the CAR) points to the object occupying that position in the list, which, among other things, can of course be another list; the second pointer (the CDR) points to another cons-cell, which in turn contains the next element in the list, and so on. A list is terminated when the NIL-pointer is in the CDR-position of the last cons-pair.

Complex objects constructed with lists are easily traversed and dynamically modified with Lisp's built-in functions. In Lisp practically everything is (or at least can be) expressed as a list so, for example, there is no difference between how data and programs are stored in the machine. This is one major reason for Lisp's power. Another is the modularity and clear layout of programs that comes from the functional (lambda calculus) basis of the language.

Lisp systems generally give the programmer a very flexible and interactive environment: functions are easily defined and tested by the user and may be separately compiled for speed; mixing of compiled with interpreted codes is possible and extensive debugging support is available. There are many different dialects of Lisp, but it seems that the dominant version for the future will be Common Lisp—an attempt to standardize Lisp.

Expert systems

In its most simplified form an expert system (or production system) consists of a data base (called a working memory) containing fundamental facts about the domain in question, a collection of rules (rule base) and finally an inference-engine (rule interpreter) that makes it possible to put queries to the system and run the rules in the data base. The inference-engine operates either by forward- or backward-chaining. Expert systems can be superimposed on Lisp or Prolog (or, as in many cases, built from the ground up as a completely separate language).

Expert systems are slow compared to more conventional languages because they often use an interpreter, which is implemented in some other language (typically Lisp, C or Pascal), the matching operation is complicated (potentially it must be performed throughout the whole working memory for each rule in the rule base for each execution step) and the number of inference steps can be large. The speed of a large OPS type expert system (2000 productions, say) running on a conventional 'supermini' is in the range of 1 to 300 production firings per second depending on implementation.

The matching operation is the critical operation. It is inherently sequential and often (because of simple instructions, pointer operations and tests in memory) limited by the available memory bandwidth. The Rete algorithm takes advantage of the fact that the contents of the working memory usually change very slowly and also that there are often many common sub-expressions in the left-hand sides of the productions which need only be evaluated once.

A promising extension to basic expert-system technology is 'blackboard architecture', in which a group of agents (expert systems) representing multiple levels of the problem domain communicate via a global 'blackboard'. An agent is triggered into action by an event on the blackboard and may eventually generate an event of its own, which in turn will cause one or more other agents to react. Difficult recognition problems in speech and sonar detection have been successfully handled with this approach.

IV. Requirements for symbolic processing

In this section we describe a collection of architectural features or techniques commonly used to enhance the performance of symbolic-processing machinery. Some of these are also found in high-speed numerically-oriented computers, others are oriented towards a specific language feature and still others increase the performance of a special architecture. Most of them can be found in some form or another in the AI-computers commercially available today.

Virtual memory

One characteristic of symbolic processing programs is that they require immense amounts of memory space. This is because of many things, including: (a) the more complex way symbolic processing languages store their data objects; (b) the fact that symbols are created and discarded dynamically during the execution; and (c) the tendency of AI programs to be large, as they often address complex problems.

For reasons of cost and technology (space and power limitations) it is often not possible to have as much physical ('real') memory in a machine as a demanding processing task requires. One way to overcome this problem (at least partly) is to use virtual memory. Programs can be run that are only partly in the physical memory, that is, the memory space used by the program can be bigger than the available memory. Virtual memory is implemented by means of an intermediate conversion table, called a page table, between the processor and the memory. This is a part of the memory management unit (MMU).

The page table maps a logical address (from the processor) to a physical address in the memory. This mapping is carried out in chunks of memory called pages. For each mapping entry in the page table, there are also a number of bits that carry information about this particular page. Among other things this information reveals whether the page is in the physical memory or stored on some secondary storage medium (e.g., a magnetic disk). If the page is not in the memory, the program is interrupted and a special page-swapper program is run instead to determine whether there is a free page available in the physical memory. If so, it reads the required page from the external storage medium into the free location. If not, the program must first decide which of the pages in the memory should be transferred to the external storage medium, in order to make room for the new page. Commonly called 'as demand' paging, this scheme means that the system never reads a page unless it is needed.

The scheme works and gives acceptable performance because of the fact that a program, although it may be large and use lots of memory for data items, over a short period of time has some locality in both the instructions executed and the data it is operating on. This is called the program's working set. If the working set is too large or the physical memory allocated to the task is too small, the system will get into a mode—trashing—in which pages are swapped back and forth between the memory and the secondary storage and very little useful work is done.

Virtual memory is also a separation between a task's (user's) logical address space and physical address space (what is on the memory bus address lines). This is convenient from other points of view; there is no need for the programmer to worry about physical addresses, programs can be run anywhere in the machine's memory, tasks can be protected from each other, and so on.

Caches and instruction-prefetch

Memory bandwidth is a very critical parameter. The performance of many processors is limited by the memory bandwidth available. In a von Neumann type of computer two streams flow between the memory and the processor: the instructions to be executed and the data being processed. The traffic to and from the memory can be reduced by the use of caches (often one cache is allocated to the data stream and one to the instruction stream).

A cache memory is a small high-speed memory that is used together with a large (necessarily) slow 'main' memory and utilizes the local character, or locality of instructions and data. Locality means that the probability of execution of the next instruction in the instruction stream is high, as is the need for access to a data-item 'near' to the one being processed. At least this is so in conventional computing. When it comes to symbolic processing things are not so rosy: in Lisp for instance there is no guarantee whatsoever that the next item in a list is located anywhere nearby in address space. There are architectural 'tricks' available to overcome this problem. The cache works by moving small chunks of data (typically 64 to 256 bits—this is called 'a line') from the large memory into the cache as needed. The data read are the data needed plus a number of adjacent data items.

When the data items are instructions, this is also called instruction 'prefetch'. In some machines there is a separate processor for prefetching instructions, which also tries to guess where branches lead and to execute other partial decoding of the instruction stream before it is fed to the main processor. Some kind of associative memory is used to store the addresses of the items currently in the cache, so that when the processor requests some data, the associative memory is searched for the corresponding address entry. If it is found, the cache supplies the data to the processor and if it is not, the cache mechanism replaces an older entry in the cache with the requested item from the main memory.

The address-matching procedures, the replacement-algorithm and the main memory operations must all be implemented in the hardware because of speed considerations. The whole idea of a cache is that it should be considerably faster than the main memory (access time: 150–1000 nanoseconds) and that often means cache read-times in the region of 30–100 ns. An entry in the cache that has been modified by some operation can be transferred back to the main memory in two different ways. In a write-through cache, the actual cell in the main memory is immediately updated when the processor is performing a write operation. This is the safest thing to do in a multiprocessor/shared memory environment because a specific memory location always corresponds to the cache belonging to a processor allowed to do updates to that cell. From a performance standpoint, write-through is not optimal because of the (very common) case of immediately updating the cell just written (for example: a variable used as a loop counter). This is of course dependent on the way the memory system and especially the cache-memory interface functions, the degree to which memory and cache operations can be overlapped, and so on.

A write-back cache does not update the corresponding memory cell until the cache-replacement algorithm decides that the positions in the cache occupied by the 'dirty' (written to) location are to be used for some other entries. So at a given instant there is *no* guarantee that the main memory is correctly updated. The benefits of this approach are a greatly reduced demand on memory bandwidth and a correspondingly increased hit rate for the cache, which in turn increases processor speed. A common technique in multiprocessor systems is to specify the 'shared' variables as non-cacheable.

Pipelining

By partitioning a complex operation into a time-ordered sequence of simpler operations and mapping each of these operations to a piece of hardware and having one hardware stage feed its results to the next in line, it is possible to have several complex operations going on concurrently albeit at different stages. The advantage is that although each complex operation still takes the same amount of time as without the pipelining, a second complex operation can be started as soon as the first computation has finished the first stage and moved on to the second stage. This interval, 'the pipeline clock', is the total time it takes to do a complete operation divided by the number of stages. In this way conventional 'super'-computers for scientific calculations achieve very high throughput for operations on vectors; by having a functional unit for floating-point addition, for example, divided up into seven segments, the time for each addition is reduced by the same factor if the length of the vector is sufficient to render the start-up time for the pipeline insignificant.

In a symbolic processor pipelining can be applied to many operations, such as tag operations and unification. Often the whole central processor unit (CPU) is partitioned into pipeline segments. There is also a software equivalent called 'stream' operation.

Tagged pointers

A common data type in symbolic processing is the pointer, which is an indirect reference to some other entity. This other entity can be a list, a composite structure, a floating point number, another pointer or something else. Often one is interested in quickly finding out what a pointer actually points to, and if this is another pointer what that in turn points to. One mechanism for achieving this is to encode what the pointer points to in a bit field, a tag, that is concatenated with the pointer itself. In machines that are not specially built for pointer-based data structures, the tag is put in the higher end of a data word that is used as a pointer. For instance, given a machine with 32-bit word length, the upper 4 bits are used as a tag and the rest as the actual address of the memory cell in question. This has two unfortunate side-effects: the addressing range of the pointer is reduced (in this case from 32 bits to 28 bits) and before actually using the pointer the tag has to be removed by some masking operation.

In a machine built for AI applications, support for handling pointers and tags is built into the address- and memory-handling hardware; the processor can immediately jump to the appropriate (macro- or micro-) code without having to first extract the tag, pointers pointing to pointers are automatically 'dereferenced' (that is, followed until an object that is not a pointer is reached). It also directs data to a suitable unit, for example when doing arithmetic with floating point numbers, the arguments are directed to the floating point unit without the programmer (or the compiler) having to explicitly put it in the code.

Stacks

Because AI-languages are often functional in their nature (explicitly like Lisp or implicitly as in Prolog) and use recursion to a high degree, it's important to have good support for stack operations. This can be implemented as a special small hardware stack located on the processor board (often called the PDL-buffer in Lisp-machines) which keeps the top items of the stack in a circular queue. A problem is the limited size of this special memory and special techniques have to be used to move the bottom of the stack to and from the main memory. For fast arithmetic (especially in Lisp) there is often a special hardware stack available for storing intermediate results in arithmetic expression evaluations.

Unification

The unification operation, which is very common in Prolog and other languages that use pattern matching, can be implemented as a special hardware unit, possibly as an add-on module for a conventional machine. Or, if the target architecture already has adequate support for dispatching on tags, unification can be implemented as a microcode routine.

For a hardware unit, the most straightforward implementation is as a small microprogrammable processor, which can efficiently dispatch (jump) to the appropriate routine in the microprogram depending on the pointer-type tag bits. Some complications are that the unit has to be recursively invoked and that many memory references are often made within one unify operation, so extensive pipelining of the unit is not easy to implement. Compared to a software implementation, a performance gain of the order of a magnitude is generally possible. It should also be noted that for Prolog, efficient compilation can in many cases reduce the number of full unifications in favour of a few simple tests. A variant is to perform a so-called 'mock-unification'—a unify operation with no variable bindings. The advantage is simpler hardware and the large number of mock unifications can be done concurrently over the whole or part of the rule base (depending on the number of units available). Because a mock unification does not change the global state, no inconsistencies can occur between different unify units. Of course a mock-unification is just what its name implies, so the purpose of the whole operation is to collect viable candidates for subsequent full unification in software and/or microcode.

Garbage-collection

Most symbolic-processing systems dynamically create new memory objects during the course of execution and perform automatic storage management, so that allocating a new object can initiate a garbage collection (GC). This is a reclaiming of objects that can no longer be referred to by any other object— that is, objects that are not useful to the system. In conventional programming languages—Fortran, C, Pascal—it is up to the programmer to explicitly request and deallocate memory space (if it is at all possible to perform this kind of operation). In symbolic-oriented languages the complexity of the memory structure is often much higher (and partly hidden from the programmer) so that explicit handling of memory storage becomes too difficult for the programmer. Some form of garbage collection is essential to every system without an infinitely large memory.

Many different algorithms have been invented for garbage collection over the years. The type of GC chosen for a system has a profound effect on performance, and although most of the time GC is implemented as a software sub-system the implications for the underlying (hardware) architecture could be great, especially for memory design, tagging and so on.

Microcode

A microcode is a list of small program steps which can be combined to perform according to a prescribed sequence. It occupies an intermediate level between software (Lisp, etc.) and the underlying hardware and is the interface between the macrocode and the electronic gates. Machines for this kind of application often allow the programmer to 'micro'-code time-consuming and/or time-critical operations. Microcode allows close control of what is going on in the processor, such as data flow and manipulation, often utilizing available processor resources and using parallelism to a degree that a higher-level language would not permit. The speed-up obtainable from this practice is in the range of five to a hundred times. The drawbacks are that it is extremely hard to code, debug and maintain microcode. It requires an intimate knowledge of the inner workings of the machine, much more so than conventional assembly programming. Some systems have a microcode compiler that more or less automatically translates high-level code to microcode.

V. Sequential processing

The uni-processors used for symbolic processing are basically conventional von Neumann type machines whose architecture is enhanced with the techniques outlined above. Some of the commercially available machines are built with surprisingly conservative hardware technology, characterized by many small- to medium-scale integration (SSI and MSI) transistor-transistor logic (TTL) chips, low system clock speeds, and the central processing unit spread out over many circuit boards. Several manufacturers are currently trying to market more high performance VLSI (very large scale integration)-based machines that will also reduce the cost of dedicated computers for AI programs.

The hardware tools for today's AI-programmers are often expensive single-user, single-processor 'work stations'. These machines are characterized by having a microprogrammed processor (or a high-performance 32-bit VLSI microprocessor); a bit-mapped display unit with typically 1024×1024 pixels resolution; a pointing device (a mouse) and a high-speed local network (Ethernet) for communication with disks, laser printers and other computers.

Lisp machines

The original Lisp machine was developed at MIT's AI laboratory by Richard Greenblatt and others in the late 1970s. It is the forerunner of many of the Lisp machines on the market today and had basically all the enhancements available to a von Neumann-type machine to adapt it for efficient Lisp-processing. Subsequent models vary in the length of a data word, the number of tag-bits, the size of the address space and the relation between what is implemented in microcode versus hardware. Physical size and power requirements have also been reduced. Although it is claimed that the architecture is general-purpose and could be adapted through microprogramming for different languages, in practice the MacLisp-derived and very rich ZetaLisp dialect is used. Other languages are compiled to Lisp and if needed complemented with microcode support for especially time-crucial or complex operations (e.g., unification).

TI Explorer

An interesting development in this line of MIT-descendant Lisp machines is the VLSI implementation of the Texas Instruments Explorer; the so-called Compact Lisp Machine (CLM). The CLM was developed with funding from the US Department of Defense and will have a big impact on the use of symbolic processing in military applications.

The CLM consists of at least four circuit boards, described in some detail below, and will eventually be housed in a military standard factor package with an approximate size of 20 by 20 by 30 cm, a weight of 10 kg and consuming just over 300 Watts of electrical power. The interconnecting back-plane bus is the 32-bit general purpose NuBus, a bus design that originated at MIT. The NuBus has a maximum transfer speed of 37.5 MBytes per second, supports multiprocessor operations, a flexible interrupt mechanism for real-time application and uses a low number of back-plane signals.

The processor is built with 2-micron CMOS technology; contains 550 000 transistors, a large number of which are used as memory cells (for dispatch memory, PDL buffer and scratch-pad memory); and will run at 40 MHz clock frequency. The chip is microprogrammable via an external writeable microprogram store of 16 K by 64 bits. The package for the chip is a pin-grid array ceramic capsulation with 264 pins.

The cache/mapper module has two parts:

1. A two-way set associative cache storing 16 K words of cached data. The cache is 'write-through' and moves data to and from the main memory in four-word chunks.

2. The mapper part handles the translation from virtual address space (128 megabytes) to NuBus address space (4 gigabytes) via two Translation Lookaside Buffers (TLBs) each containing 2 K of page mapping. To support operation in a multiprocessor environment, parts of the virtual address space can be declared as non-cacheable. This makes it possible to maintain consistency in shared memory areas. A memory board built up with 256 K byte dynamic memory chips has a capacity of 2 megabytes. The board does parity checking on the byte level and supports four-word block transfers.

Finally, the Multibus Interface Module hooks together the industry standard, but rather slow, Multibus I bus with the high-speed NuBus doing appropriate mapping of addresses, interrupts and related operations.

The CLM is object code compatible with the TI Explorer but will be two to three times faster. Initially software development and interactive debugging is done with the Explorer as host. A Common Lisp environment is under development.

Xerox Lisp machines

Another family of Lisp machines—also conventional single-processor machines but of a quite different flavour—was developed at the Xerox Palo Alto Research Center (PARC) from the early 1970s. These machines are for executing, among other languages, Interlisp, a different dialect of Lisp, and use an architecture that is more oriented towards vertical microcode, and narrower data paths and is realized with higher-speed IC-technology.

Prolog machines

Unlike machines for Lisp, machines specialized for running Prolog programs are not yet widely available. Apart from experimental laboratory machines, only the Japanese have attempted to build a machine that could be produced in large numbers.

On the other hand, there are many software-based Prolog systems available for a large variety of computers. Several of these, especially those aimed at the different Lisp machines, have extensive microcode support. There is work going on in building back-end Prolog processors for conventional work stations, and so on.

The PSI-machine

In the early 1980s, researchers at ICOT, the centre for the Japanese Fifth Generation Computer Project, developed a single-processor machine oriented towards Prolog execution: the Personal Sequential Inference-machine (PSI). The PSI has all the conventional outer attributes of an AI work station: high-resolution bit-mapped display, a mouse, local network interface and moderate physical size. Architecture-wise it is a microcoded tagged-pointer machine with a 40-bit word size, 8 bits for tags and 32 bits for data, and several special features for Prolog execution:

1. An 8-K word write-back cache specially adapted to stack operations.
2. A 1-K word register file accessible by different addressing modes (direct, indirect, frame-oriented, etc.) including facilities for certain space-optimizations during execution.
3. Microprogram tag dispatching.
4. Hardware support for up to 63 concurrent processes. The logical address space is 16 M words and there can be up to 256 such 'areas' or memory domains.

5. The machine language is Prolog derived. The machine instructions are high-level, in fact the PSI executes Prolog by way of a microcoded interpreter.

The PSI-machine is built with standard Schottky TTL logic components and high-density MOS RAMs, the clock cycle is moderately high. The whole machine with 16 M words of memory and assorted input/output devices occupies circa 40 medium-size circuit boards.

The Warren Abstract Machine

David Warren, a pioneer in the area of compilation of Prolog programs, has specified a 'virtual' machine for efficient execution of compiled Prolog in both space and time, the Warren Abstract Machine (WAM). (A virtual or abstract machine has hypothetical computer architecture—in other words an imaginary instruction set and its semantics. Usually virtual machines are emulated in software, or microcode, on the various target machines.)

Surprisingly, the WAM is based on conventional stack and heap-based architecture, much like the optimal architecture for conventional languages such as C or Pascal. However, the bulk of the WAM instructions deal with the handling of Prolog variables and the proper invocation of the different clauses. The instruction set can be roughly divided into five different groups:

1 and 2. Get and Put-instructions—responsible for the transfer and matching (unification) of arguments during clause invocation.

3. Unify-instructions—which unify structures (complex data objects) and create new structures.

4. Procedural-instructions—for control flow and handling of environments.

5. Indexing-instructions—which select the correct clause for execution.

There are also a number of registers, and three stacks: (*a*) the environment and choice-point stack (saves local variables and state of the computation); (*b*) the heap stack (contains structures and lists built during unification and clause invocation); and (*c*) the trail stack (keeps a record of variable bindings that must be undone on backtracking). The WAM has attracted great interest among Prolog researchers and practically all compiler developments are targeted for the WAM or variants thereof. There are also several WAM-related hardware projects.

Some interesting questions are raised by the WAM-developments concerning the efficiency possible for compiled Prolog and the kind of specialized hardware support which is really needed for the processor. As it turns out, Prolog can be very efficiently compiled for the conventional 32-bit VLSI microprocessors available today. A high-performance processor with simple fast instructions, powerful bit-field operations for tagged-pointer operations and a large flexible register file are perhaps all that are needed for fast execution of Prolog. Several research projects are also investigating the adaptation of WAM for parallel execution models.

VI. Parallel processing

Introduction

Parallel processing is the most fundamental change to occur in the computer industry to date. Electronic signals travel approximately 30 cm in 1 nanosecond. Thus, a computer with a 1 ns cycle time is under severe space constraints. Today's supercomputers have clock cycles (not necessarily the same as cycle time) in the region of 10 nanoseconds. As can be seen there is a limit to how much can be gained in performance by increasing the clock speed. This is the 'speed-of-light barrier' in computer architecture.

The only way to radically increase the performance of computers built with conventional component technology is to employ massive parallelism, that is, to have a large number of processors work on different parts of the same problem. Unfortunately this is not easily accomplished. There are fundamental problems concerning the architectural structure of a parallel machine and its programming.

The assumption that lies behind much development of multiprocessors is that by increasing the number of processor nodes there will be a more or less proportional increase in the performance of the system, and that this can be achieved in a way that is transparent to the programmer. This is not necessarily so. In a naively designed system there may well be an initial linear increase of speed but, as the number of processors exceeds 10 to 20, there is actually a decrease in performance! The reasons for this are almost invariably sharply rising communication overhead, because of contention for globally-shared information and a suboptimal distribution of the workload between participating processors. These are the two major problems with multiprocessor systems.

Often there is also a requirement for the programmer to annotate or partition the program to be able to achieve an acceptable performance. This is a very hard thing to do and research is under way on how to do an automatic static analysis when compiling a program in order to extract its possible parallelism. Assignment and changes to (global) state are troublesome when it comes to achieving parallelism because of the problem of keeping the coherency of the involved processor nodes.

Global versus local memory

Multiprocessors can have different types of memory architecture:

1. A global memory that is shared between all the processors. This means that a processor has to contend with other processors for the use of the memory. If this is a frequent operation, it is a severe bottleneck.

2. With local memory the processor has full access with no contention, but interprocessor communication must still go via the shared memory. There

is often a special communication network built into the hardware to use instead.

3. The so-called 'dance hall' model which has a number of processors and a number of memory units. The processors can be individually connected through a special network to one or more memory units. The drawback is the often large delays introduced by the network when accessing the memory. A cache memory on the processor side of the network can reduce the otherwise inevitable performance degradation.

Interconnection networks

A very important component in a multiprocessor system is the interconnection network (ICN). Characteristics of this network, such as topology, routing algorithms, throughput, latency and reconfigurability greatly influence the design of the rest of the system.

Packet versus circuit-switched network

Two principal ways of setting up a connection between two points are circuit switching and packet switching. In a circuit-switched network a direct 'physical wire' channel is set up between the two ends before the actual transmission begins. This means that all intervening nodes or stations must allocate a path, each one a step on the way to the intended destination. When the end node is reached and the allocation is ready, this has to be acknowledged back to the sender. The time to set this up can be considerable for a large network. The channel is maintained throughout the transmission and does not depend upon whether any data are actually sent. There is no congestion once the channel is set up and transmission time for a message passing through it can be guaranteed. One important feature is that the sequence of the individual data items is kept in a circuit-switched net. The drawback is that for low-intensity traffic it is a waste of expensive network bandwidth and available channels.

The packet-switched network works by accepting a small packet of data from a node (the sender). It then reads the destination address encoded in the packet and tries to decide upon a suitable intermediate node to forward it to, in order to get the packet closer to its destination. If no path can be found because the neighbouring nodes are busy, the packet is temporarily stored and a new attempt to forward it is made at a later time. This process of 'node-hopping' is repeated until the packet has reached its destination.

This scheme has several implications. The packets cannot be kept in order; since a packet may overrun an earlier packet, some kind of sequence number has to be used. The transfer times cannot be determined beforehand, so the nodes have to be able to adapt to this varying latency. Of course a statistical average throughput can be computed for the network. Sometimes the preferred path through the net is given in the packet itself.

The advantages of a packet-switched network are the dynamic routing of messages; if one path is blocked another can be selected, and thus efficient use can be made of available channels.

Some simple topologies

In the crossbar topology every node is directly connected to all other nodes. This is a simple and powerful scheme and the complexity of the interconnections very quickly becomes unmanageable. It is nevertheless very useful for a low number of nodes. Fixed-array structures or grid-like layouts suffer from inflexibility and it is easy for isolated nodes to suffer from starvation (no work). But many physics-related problems map nicely on to grids.

The perfect shuffle network in its simplest single-layer version has a butterfly-like pattern through which a message is routed in multiple passes, going through a number of intermediate nodes, on its way from the sender to the destination. A hypercube is a generalized n-dimensional cube. One convenient property is the existence of simple algorithms to route messages utilizing redundant paths. It is actually a member of the family of shuffle networks. Other possible configurations are the ring bus and the tree-structure which suffer from a severe bottleneck at the root.

The problem of synchronizing a large high-speed ICN is not a trivial one. Should it have a global clock, that is, a synchronous net, with its attendant clock skew and other distance-related degradations? Or should it be asynchronous, that is, with the clock extracted from the data stream, with the added complexity this adds to the circuitry and problems with meta-stable states, and so on?

What level of granularity?

Granularity denotes the level of decomposition of a problem into sub-problems. Used in this context it is the size of the work unit allocated to each processor, when a program is mapped on to a multiprocessor. Fine granularity would mean work units of the size of machine instructions or even lower, a large or coarse granularity indicates size on the level of complete routines or programs. The chosen level of granularity affects the time of communicating with other nodes, the structure of the ICN, how well load balancing can be done and the cost of context switching.

A fine-granularity machine has a larger need for communication between its processing nodes, because each node does relatively little and thus has to communicate its results to other nodes and receive new jobs quite often. A problem is that there seems to be a more or less fixed time (overhead) required for the transmission of small chunks of data through an interconnection network, so at some grain-size this overhead completely overshadows the execution time for a job in the node.

The finer the granularity the more effective the load balancing that can be done. Consider for example the level of granularity of complete subroutines. If a couple of the routines are significantly larger than the others, it is obviously difficult to get an equal workload even on a small number of processors.

Parallelism in logic programming

Basically there are two forms of parallelism available in Prolog programs:

1. OR parallelism can be utilized when a sub-goal of a query matches the head of more than one clause (rule). It is the search for alternative solutions to a problem. The difficulty of OR parallelism lies in the handling of the multiple variable bindings emanating from the different and independent branches of the search tree. A sub-form of OR parallelism is search parallelism, which is a parallel search for all clauses whose heads match the sub-goal.

2. AND parallelism permits at least two sub-goals within a goal to be solved in parallel. In other words it is the parallel investigation of sub-problems within a larger problem. The trouble occurs when those sub-goals share one or more variables. The synchronization of the bindings has to be maintained, that is, the binding and unbinding of variables has to be done in the proper sequence and propagated to all processes participating in the execution. Stream parallelism is a variant of AND parallelism, where the sub-goals have a producer-consumer relationship to each other.

Parallelism in expert systems

In an expert system parallelism can be utilized at three different levels:

Production level

Each rule (or group of rules) is matched in parallel. This has the advantage that no communication between the different processors is needed during the actual matching, but updates of the working memory have to be communicated between all processors. There is a problem of low utilization of processors that comes when the initial matching phase is over and only some of the processors (those whose match operation succeeded) are doing useful work.

Condition level

Here each condition in the IF-part (the left-hand side of the rule) is matched in a separate processor and the results are combined. This implies increased communication overhead which, together with the fact that the condition part tends to be simple (without many conditions), means that this parallelism is of limited use.

Action level

When a rule has been triggered, the changes that are to be made to the working memory are carried out in parallel. There is not much parallelism to be gained from this level, because of the usually small number of changes to working memory required and also the imposed bottleneck of the global working memory.

The increase in performance that comes from using the parallelism of conventional rule-based expert systems is limited to a factor of about 20, depending on the size and structure of the problem domain.

SIMD versus MIMD

Two important terms that are used in connection with multiprocessors are Single Instruction Multiple Data (SIMD) and Multiple Instruction Multiple Data (MIMD):

1. A SIMD machine has a single centralized instruction store from which instructions are distributed to the (many) individual data processors. The data processors all execute the same instruction at a given time and the program counter is global, although there is often a facility for conditionally inhibiting the execution of an instruction on a local data processor depending on the result of a previous operation.

2. In a MIMD machine each processor has its own local instruction store along with its data-processing functions. This means that any one processor can execute a completely separate program from any other (or, as is perhaps more common, different parts of the same program). The possibility of a SIMD machine conditionally executing instructions in a data-dependent way makes it possible for a SIMD machine to emulate a MIMD machine, albeit with reduced efficiency.

Some parallel machines

A class of shared-memory machines

There are a number of commercially available parallel machines. The most successful ones use rather simple architectures. The basic concept is as follows: A number of processors, usually less than 30, are all connected to a global bus. This bus is a more or less conventional computer-bus with a moderately high bandwidth, often enhanced with the ability to perform block transfers of data and interleaving of data requests and responses from multiple clients. The memory is also connected to the bus that constitutes the globally available memory pool. A processor board typically contains a 32-bit VLSI microprocessor, a memory management unit, a cache and/or a small local memory. Usually there is also some other means, often a secondary (serial) bus, for the processor nodes to communicate with each other. This special channel is used to arbitrate between multiple processors wanting to access the same global resource and also for the distribution of the workload between the nodes.

The processors execute programs out of the global memory, where data are also stored. Information can be communicated between nodes either by sending messages through some kind of 'mailbox' in the global memory, or by sharing blocks of memory. Contention for such shared resources is in the form of critical sections protected by spin locks, where one processor at a time can hold a lock, and all others trying to get access wait in a loop, testing the lock. This is called busy waiting.

Non-shared data and the execution state of each processor can be stored in the local memory. The granularity of the parallelism is coarse, typically at the level of a user program or process. This means that there is not much communication between different nodes or context-switching on each node. This class of machine offers relatively high performance at low cost, and often successfully replaces larger uniprocessors. They are the result of well-balanced engineering decisions regarding memory and bus-bandwidth versus the speed of available microprocessors.

Multiprocessors of this type will not offer a massive increase in computational power, largely because of the bottleneck introduced by the global bus and memory. For the same reason, it is not possible to scale this kind of architecture to thousands of nodes. Refinements in the form of sophisticated cache schemes can only partly alleviate this. Still, these machines offer an amenable environment for experimenting with parallel-execution models, and there are several implementations of parallel symbolic processing systems going on world-wide.

The Cosmic Cube

The Cosmic Cube is an experimental multiprocessor designed and built at Caltech by a group led by Professor Charles Seitz. A modified and expanded version, the Personal Scientific Computer, is marketed commercially by the Intel Corporation. The original Cosmic Cube consists of 64 processor nodes, each containing a 16-bit microprocessor and associated memory and, important for the type of application for which the Cube was originally aimed, a floating point co-processor.

A node is connected by a bi-directional point-to-point communication channel to six other nodes which in turn are connected to other nodes, and so on. The topology of the interconnections can be viewed as a cube in six dimensions where each interconnection is an edge, this is called a hypercube. The nodes communicate by sending packets of information to each other, if a node is not in the immediate vicinity the packet is routed through intermediary nodes.

The BC-machine

A scheme for OR-parallel execution of Prolog-programs on small- to medium-scale multiprocessor systems has been proposed by K. A. M. Ali and L. E. Fahlén and a prototype is under construction at the Swedish Institute of Computer Science (SICS). The machine is called the BC-machine, where BC stands for BroadCasting, which is a central feature in the architecture.

The basic idea is as follows: assume a set of processing elements each with local memory for storage of codes and data (called a PEM); all the PEMs are interconnected by a special communication network (ICN) and there is also a processor manager (PM) which has some means of controlling the PEMs (i.e., start, stop, interrupt . . .) and setting up the ICN (by which a PEM communicates with one or more other PEMs). The PEMs can also communicate with the PM, that is, send a message, typically one-word long.

To run a program on the system all the PEMs are loaded with identical copies of the (Prolog) code to be executed, one PEM is assigned to be a master and the rest to be slaves. The master starts executing the root branch of the program and when doing so broadcasts the changes it makes to its state (the Prolog execution environment, the stack, heap, reset list and so on) to all the other PEMs connected via the ICN. This continues until the master-PEM in its progress through the program code comes across a suitable choice point in the execution graph; the master then sends a message to the PM (the manager).

When the PM receives this message it initiates a split operation, which results in the establishment of a second master with its own slaves. The new master and associated slaves are former slaves to the first master and are therefore in an identical state to the first master and its remaining slaves at this point in time. If for simplicity's sake we assume that the choice point only has two choices (OR-parallel branches) the execution continues by the first master going down the left branch and the second master taking care of the right branch.

The above process is repeated on the sub-trees of the execution graph that the two branches represent. As long as there are enough processors available to cover all branches at every split point the execution can proceed with a very high degree of parallelism. There will be very little overhead when performing a split as all updating of environments in the slaves is done via the broadcast mechanism during the course of execution between the split points. That is, there is no explicit copying of environments at the actual split point.

A more realistic situation is of course one in which processors are a scarce resource. Then, when a branch fails the affected PEMs have to be reallocated as masters or slaves on a completely different branch on the search-tree. The problem here is that the reallocated PEMs have a partially incorrect state in regard to the new branch. How much the environments differ depends on the distance between the actual points in the execution-graph. The whole or part of the environment resident in the master on the target-branch must be copied to the PEMs that are to be reused (allocated to that branch). There are of course several problems and further implications of the scheme; this description is simply to introduce the execution model of the BC-machine.

The Connection Machine

The Connection Machine is a SIMD processor with extremely fine granularity. The processor elements, of which there are 65 536 in the present model, are ultra-simple. They are 1-bit processors with 4 K bits of local memory. Sixteen of these processors and a network router are integrated on one VLSI chip. The memory is in four external RAM chips connected to each processor chip. In total there are 32 megabytes of memory. The 4096 processor chips are hardware-wise interconnected in a hypercube topology. Virtual channels between any processor elements are easily set up in software.

The dynamic reconfiguration possible via virtual channels together with the large number of processors gives the machine the power to adapt itself to the structure of the data in a problem. A central idea in the design of the Connection Machine is that the cost should be in the network. This comes from the insight that the major obstacle in a parallel processor is the scarcity of communication bandwidth between the nodes. A C compiler and a Lisp compiler have been developed for the machine.

Several applications have been successfully demonstrated on the Connection Machine: text retrieval from a data base, VLSI-chip simulation, fluid-dynamics and stereo-image matching.

DADO

There are also parallel machines designed to execute rule-based expert systems. An example is the DADO machine from Columbia University, which consists of a large number of simple processors, each with a small local memory, connected in a binary-tree topology. The DADO has an interesting hierarchy of processor sub-trees executing in MIMD or SIMD mode, with the working memory elements stored in the leaves of the binary tree, the rules each stored in a processor node at the middle level and the control regime residing in the processors at the root level. Changes to the working memory are broadcast from the top of the tree in a very efficient manner.

Signal processing

In many AI applications, especially in the areas of speech, vision, sonar and so on, the symbolic processor can have a front end consisting of several digital signal processors (DSPs) that conduct massive number-crunching on the incoming data from the sensors. This can for instance involve Fast Fourier Transforms (FFTs), correlation and matrix manipulations. The objective is to perform extraction, reduction or abstraction operations on the data to put them in a more usable form before they are input to the inference engine.

Commercially available low-cost VLSI signal processors can typically perform 5 to 10 million instructions per second (MIPS) and more exclusive building blocks for a DSP can be used to achieve 25 million floating-point operations per second (MFLOPS). There are also very high-performance specialized VLSI processors for performing filtering, FFTs, convolution, encoding/decoding, compression, and so on. In this context it is interesting to note that the add-on arithmetic 'co-processors' that are available for today's popular 16- and 32-bit microprocessors are generally not more powerful than the floating-point units supplied with today's (or yesterday's) larger minicomputers (e.g., VAX). The performance is in the region of 0.1 to 0.5 MFLOPS.

VII. Developments in VLSI and related areas

Gallium arsenide (GaAs)

This material has three major advantages: (*a*) higher speed, an order of magnitude greater than TTL. It is not impossible to have a clock frequency in the region of 2–5 GHz. Speed comes from the higher electron mobility and lower parasitic capacitances of GaAs; (*b*) it has a lower power consumption, an order of magnitude lower than ECL; and (*c*) it is *less sensitive to environmental conditions*, tolerating radiation levels in the range of 10–100 million rads and temperatures from −200°C to 200°C. This is of course a point that is very important for military applications.

The material also has several drawbacks, however. GaAs components are one to two orders more expensive than similar silicon designs, because of more expensive raw material and much lower yield. This will improve in the future because of better manufacturing processes. There are similarities in the process technology used for GaAs and for conventional silicon that could improve 'the learning curve' for GaAs. The yield problem could also be dealt with at the design level by incorporating fault-tolerant circuitry.

The chip density is at least one order of magnitude lower than for CMOS because of yield and power considerations. As an example: a state of the art GaAs may contain 30 K transistors and consume 2 W of power; a commercial CMOS part like the Motorola M68020, a 32-bit microprocessor, has 200 K transistors and the same power consumption.

To utilize GaAs in an efficient way it is not appropriate to use existing MOS designs. The architectural implications of the following must be considered:

1. The transistor count is lower so simpler processor designs are preferred.
2. The ratio between on-chip/off-chip gate delays is much higher, so it is expensive performance-wise to access memory or registers off-chip.
3. This problem is made more difficult by the fact that there are not enough resources available on-chip to be able to have a cache and a large register file.
4. The amount of intra-chip connections is limited by the low gate-in/gate-out capacity (how many other gates one gate can comfortably drive).

A typical trade-off for a processor built with GaAs would be to have a bit-serial multiplier instead of the usual parallel-adder/microcode combination. Another implication is that microwave technology should be used for off-chip communications. The speeds obtainable for arithmetic components are on a par with the scalar performance of present-day super-computers. Some of the GaAs problems described above are in fact related to more general fundamental problems afflicting any super high-speed VLSI technology; GaAs applications just happen to be a first example.

Perhaps the most appropriate area for GaAs solutions is in custom-designed chips, rather than in standard components. It should also be mentioned that there are important non-digital uses for GaAs, especially for microwave and fibre-optic applications. Components available are, for example: 1–4 K GaAs static RAMS with access times 1.5–3.0 ns, gate arrays with 2000 gates, standard logic-gates and multiplexors.

BiMOS

A technique that has attracted some interest is the possibility of mixing bipolar and MOS technology, BiMOS, and thereby trying to combine the best of both worlds: from CMOS, the low-power dissipation and high-noise immunity and from bipolar the high drive capacity, especially for capacitive loads (bipolar technology suffers from too high power dissipation to be used in really high-integration chips). By using bipolar transistors for both intra-chip and off-chip drivers low interconnection delays are obtained combined with low power dissipation. The difficulties are the conflicting manufacturing process requirements of the two technologies.

VHSIC

The Very High Speed Integrated Circuits (VHSIC) programme is initiated and sponsored by the US Department of Defense. Its aim is to develop and manufacture integrated circuits of a complexity and speed not hitherto seen. The programme is split up into different phases and works by awarding contracts to interested IC-manufacturers. The current programme is VHSIC Phase 2, which will use line geometries of 0.5 micron, and it is projected to produce working chips by the end of the 1980s.

The chips are built with different types of processor (bipolar, GaAs and for the most dense, CMOS) depending on the manufacturer and the purpose of the chip. These superchips are up to nearly 5 cm square, compared to the 1 cm square maximum for the most advanced commercial chips available today. They contain up to 30 million devices and will run with clock speeds approaching 100 MHz. Apart from their sheer density and speed a number of important and novel features will be incorporated, with the goal of increasing manufacturing yields (a very large problem for chips of this size) and chip reliability thereby reducing the cost and expected lifetime.

Software configurable interconnections between the different functional units on the chips, in combination with extensive self-testing and error-checking and the inclusion of redundant components on the chip (that can be switched in to replace a malfunctioning unit), will enable automatic 'self-healing' of the chips from damages in the manufacturing process or from some other cause. Another possibility is to reconfigure the chips in a 'mission-specific' way. Among the types of ICs constructed are, apart from the conventional processors and memory-chips, several other more specialized components for signal-processing applications, such as floating-point units,

Fast Fourier Transformers, convolution and correlation units. Processors aimed at pattern matching and associative searching are also planned.

Optical computers

One possible future way of building computers which is under investigation is to use photons instead of electrons as the information carriers, that is, to create an optical computer.

There are several advantages to be gained:

1. Using photons as information carriers has a great speed advantage compared to electrons.
2. Light is more or less immune to cross talk and electromagnetic interference.
3. Interconnections between units can be made much more flexible, there is no need for wires, as free space will be adequate. Also this makes totally different multiprocessor architectures possible.

One major obstacle to the building of a 'true' digital computer based on optical elements is the absence of a switching element as flexible as the transistor. Another is the difficulty of integrating many optical components on a chip. The real optical computer lies perhaps 10 to 20 years in the future.

But there are other important computer-related applications for optical technology, such as the building of large high-speed crossbar switches for hitherto impossible multiprocessor architectures and their use in analog signal processing for radar applications, image processing, and so on. In the USA DARPA is sponsoring projects in these areas. Optical associative memories and holographic storage are interesting for the artificial intelligence domain.

The Reduced Instruction Set Computer

The Reduced Instruction Set Computer (RISC) is a re-examination of the trade-offs involved in resource allocation when designing VLSI processors. The concept evolved simultaneously in several computer research laboratories in the late 1970s. This was a consequence of the realization that the overwhelming part of the instructions executed by a computer were simple ones like 'add' and 'move'. Complex instructions for string-handling, and so on, were rarely used. Still, implementation of complex instructions in the machine used the major part of a processor's hardware and/or microcode and consumed a large part of the design effort. Perhaps most important was that the presence of complex instructions actually slowed down the performance of the simpler ones!

A simpler machine would be able to run with a faster clock and execute instructions in fewer machine cycles. Complex instructions could be emulated by a sequence of simple ones. And hardware resources thus freed could instead be allocated to large register banks, cache-memories, and so on. Another (perhaps surprising) consequence of having a very simple, register-oriented machine was that it proved much easier to construct optimizing compilers for such a machine.

An RISC is characterized by: (a) simple instructions; (b) single machine-cycle execution for (nearly) all instructions; (c) register operations only, with explicit memory load and store instructions; (d) single-length instruction format, with few format variations; (e) a large number of registers; and (f) a cache.

One way of summarizing the RISC concept is to say that the complexity has moved from the hardware architecture to the compiler software. The role of the RISC in symbolic processing is under investigation.

VIII. Concluding remarks

AI applications require very large memory space, and high processing power. Also, developing software for AI needs a more interactive and flexible programming environment with, among other things, a much higher communication bandwidth between the computer and the programmer.

Symbolic processing dominates most AI applications; Prolog and Lisp are programming languages, used for implementing expert systems, natural-language understanding systems and many other applications. The general features of these languages and various hardware developments are described above. Two interesting areas in computer development are not considered in this paper: object-oriented programming languages (and architectural support) and data-flow machines.

For uni-processors future performance increases will come from the combination of very sophisticated compilation techniques and simpler and faster processors with lots of local on-chip storage. One promising area is the development of a Reduced Instruction Set Computer (RISC) for symbolic processing. In such a design the special symbolic-processing requirements (mainly the handling of tags and pointers) could be implemented as extensions—'amplifiers'—to the basic simple instruction set. The code to implement very complicated functions (e.g., unification) may be stored in an on-chip 'subroutine-memory'. Another possibility is to have a special co-processor.

Although parallel execution is hailed as a panacea to the problem of insufficient processing power, there remains a large body of unsolved problems in hardware-architecture, programming languages and suitable algorithms. Many schemes work for a small-to-medium number of processors, but are more or less impossible to scale to thousands of processing elements. Very few of the existing languages are suitable for parallel programming. Often they need extensive annotation by the programmer or put unreasonable requirements on the target architecture regarding communication bandwidth and interconnectivity. Also there is a considerable overhead of running time from context switching and environment management. Lots of work needs to be done and these issues are very active research areas.

It seems that some of the key issues to achieve high performance on multiprocessors are: (a) having a very high-speed interconnection network; (b) every processor must have lots of local memory; (c) no (or very little) reliance on global (shared) resources; (d) cacheing and copying of information to be done as much as possible (re-use of information); (e) it should be possible to dynamically reallocate processor nodes to different parts of a problem as the need arises; and (f) the execution model has to tolerate long latencies, because there are always going to be very long delays involved in accessing information over a large ICN.

GaAs technology will not have a large impact on symbolic-processing computers in the near future because of the limitations imposed by the low level of integration offered. But at least one super-computer manufacturer has revealed its intentions to build its next-generation machine with GaAs components. The possibilities offered by ultra high-density VLSI-chips with tens of millions of transistors have not yet been fully realized.

Although very little is known about the inner workings of the brain, results from neuro-science regarding the neural network and the interconnectivity between neurons in the brain has inspired researchers in computer science and other related domains to think about very different computer architectures. The development of massively parallel machines also makes possible the realization of new artificial intelligence paradigms. One is the field of associative networks in which knowledge is expressed as activity spreading among simple nodes of a network and information is encoded by varying the strengths of the interconnections between different nodes. Storage is distributed and there is no distinction between memory and computation. Systems of this type may have important applications in vision and other pattern-matching and constraint-oriented applications.

Acknowledgement

The help and advice received from Dr K. A. M. Ali and Mr Dan Sahlin, both at the Swedish Institute of Computer Science, is gratefully acknowledged.

Further reading

Ali, K., *OR-Parallel Execution of Prolog on a Multi-Sequential Machine*, SICS, Stockholm, Research Report R86006, July 1986.

Ali, K., 'Pool machine: a multiprocessor architecture for OR-parallel execution of logic programs', TRITA-CS-8603 (RIT: Stockholm, Oct. 1985).

Allen, J., 'Computer architecture for digital signal processing', *Proceedings of the IEEE*, vol. 73, no. 5 (May 1985).

Altman, A., 'RISC-taking in symbolic processors', *TI Engineering Journal*, Jan.–Feb. 1986.

Amundsen, M. et al., 'Compact LISP machine', *TI Engineering Journal*, Jan.–Feb. 1986.

Bawden, A. et al., 'The Lisp Machine', *Artificial Intelligence: An MIT Perspective, Vol. 2* (MIT Press: Cambridge, MA, 1979).

Bell, T. E., 'Optical computing: a field in flux', *IEEE Spectrum*, Aug. 1986.
Bratko, I., *Prolog Programming for Artificial Intelligence* (Addison-Wesley: Reading, MA, 1986).
Brownston, L., Farrell, R., Kant, E. and Martin, N., *Programming Expert Systems in OPS5: An Introduction to Rule-Based Programming* (Addison-Wesley: Reading, MA, 1985).
Clark, D. W., Lampson, B. W. and Pier, K. A., 'The memory system of a high-performance personal computer', *Transactions on Computers*, IEEE, vol. C-30, no. 10 (Oct. 1981).
Cole, B. C., 'GaAs LSI goes commercial', *Electronics*, 18 Sep. 1986.
Conery, J. S. and Kibler, D. F., 'AND parallelism and nondeterminism in logic programs', *New Generation Computing*, vol. 3 (1985).
Deering, M. F., 'Architectures for AI', *Byte*, Apr. 1985.
'DSP boards help tackle a tough class of AI tasks', *Electronics*, 21 Aug. 1986.
Explorer Technical Summary (Texas Instruments: Dallas, TX, 1985).
Feldman, J. A., 'Connections', *Byte*, Apr. 1985.
Fura, D., Helbig, W. and Milutinovic, V., 'An introduction to GaAs microprocessor architecture for VLSI', *IEEE Computer*, Mar. 1986.
Gabriel, J., Lindholm, T., Lusk, E. and Overbeek, R. A., A Tutorial on the Warren Abstract Machine for Computational Logic, ANL-84-84, Argonne National Laboratory, IL, June 1985.
Gabriel, R. P., *Performance and Evaluation of Lisp Systems* (The MIT Press: Cambridge, MA, 1985).
Halstead, R. H., Jr, 'Parallel symbolic computing', *IEEE Computer*, Aug. 1986.
Hillis, W. D., *The Connection Machine* (MIT Press: Cambridge, MA, 1985).
Hinton, G. E., 'Learning in parallel networks', *Byte*, Apr. 1985.
Hirsch, A., 'Tagged architecture supports symbolic processing', *Computer Design*, 1 June 1984.
Hopfield, J. J. and Tank, W., 'Computing with neural circuits: a model', *Science* (8 Aug. 1986).
Katevenis, M. G. H., *Reduced Instruction Set Computer Architectures for VLSI* (MIT Press: Cambridge, MA, 1985).
Kogge, P. M., *The Architecture of Pipelined Computers* (McGraw-Hill: New York, 1981).
Kohonen, T., *Self-Organization and Associative Memory* (Springer-Verlag: Berlin, 1984).
Lampson, B. W. and Pier, K. A., 'A processor for a high-performance personal computer', *Proceedings of the 7th Symposium on Computer Architecture* (IEEE: Washington, DC, May 1980).
Lansner, A., *Investigations into the Pattern Processing Capabilities of Associative Nets*, TRITA-NA-8601 (RIT: Stockholm, 1986).
Lieberman, H. and Hewitt, C., 'A real-time garbage collector based on the lifetimes of objects', *Communications of the ACM*, vol. 26, no. 6 (June 1983).
Moon, D. A., 'Architecture of the Symbolics 3600', *The Proceedings of the 12th Annual International Symposium on Computer Architecture* (IEEE: Washington, DC, 1985).
Patterson, D. A., 'Reduced instruction set computers', *Communications of the ACM*, vol. 28, no. 1 (Jan. 1985).
Penny, N. H., 'Blackboard systems: the blackboard model of problemsolving and the

evolution of blackboard architectures', *The AI Magazine*, vol. 7, no. 2 (Summer 1986).

Penny, N. H., 'Blackboard systems: the blackboard application systems, blackboard systems from a knowledge engineering perspective', *The AI Magazine*, vol. 7, no. 3 (Aug. 1986).

'Scheme-79—Lisp on a Chip', *IEEE Computer*, July 1981.

Sohma, V. *et al.*, 'A new parallel inference mechanism based on sequential processing', *Proceedings of the Working Conference on Fifth Generation Computer Architecture*, Manchester, July 1984.

Steele, G., Jr, *Common LISP—The Language* (Digital Press: Burlington, MA, 1984).

Steele, G. L., Jr and Sussman, G. J., 'Design of a Lisp-based microprocessor', *Communications of the ACM*, vol. 23, no. 11, Nov. 1980.

Stolfo, S. J. and Miranker, D. P., 'DADO: a parallel processor for expert systems', *The Proceedings of the International Conference on Parallel Processing* (IEEE: Washington, DC, 1984).

Taki, K. *et al.*, 'Hardware design and implementation of the Personal Sequential Inference Machine (PSI)', *The Proceedings of the International Conference on Fifth-Generation Computer Systems*, ICOT, 1984.

Tick, E. and Warren, D. H. D., 'Towards a pipelined Prolog processor', *New Generation Computing*, vol. 2 (1984).

'TRW's Superchip passes first milestone', *Electronics*, 10 July 1986.

Ungar, D., *et al.*, *Architecture of SOAR: smalltalk on a RISC*, *The Proceedings of the 11th Annual International Symposium on Computer Architecture* (IEEE: Washington, DC, 1984).

Winston, P. and Horn, B., *LISP* (2nd edn.) (Addison-Wesley: Reading, MA, 1985).

Chapter 4. Comparison of human and machine intelligence in the context of conflict

ALEX M. ANDREW

I. Natural and machine intelligence

The term *artificial intelligence*, usually abbreviated to AI, was introduced in the late 1950s to denote studies aimed at producing artefacts (usually, digital computer programs) showing behaviour which would be accepted as 'intelligent' if displayed by a human. This definition is clearly not rigorous, especially as there is no universally accepted definition of 'intelligence'. Nevertheless a set of techniques, and areas of application, have come to be commonly subsumed under the heading of AI.

The subject area has intrinsic fascination and has attracted many able minds. Its aims are such that it frequently provides the incentive and the test bed for new programming techniques. It has been referred to as the 'department of clever tricks' of computer science.[1]

AI studies have been undertaken with various motivations. Some workers try to model human mental processes, in the hope of obtaining greater understanding of natural intelligence. Others are simply interested in achieving certain kinds of overall performance, irrespective of whether the method implemented in the program has any correspondence to the working of the brain. It can happen, of course, that work in AI throws light on human intelligence even when it was not intended that it should, since it inevitably gives new insight into the nature of the 'intelligent' task to which it is directed.

There is, however, a wide spread of opinion concerning the achievements of AI studies. Some sceptics dismiss the programs as a collection of 'gimmicks' providing some sort of fortuitous and superficial correspondence to the intelligent behaviour of humans. Others believe that the modelling of human intellect in AI programs should become the main technique of psychological research. The viewpoint of the present writer is an intermediate one. There is good reason to think that many of the techniques employed in AI programs (the use of *heuristics*, for example), are also features of human thought. To some extent, therefore, AI work usefully models human mental processes, but there is also much about human thinking that remains mysterious.

The forms of intelligence shown by artefacts tend to be somewhat cold and unemotional. Programs have been written for such highly regarded activities as playing chess, proving mathematical theorems and making medical diagnoses. However, machines have yet to achieve certain subtle discriminations that are readily made by animals or by humans judged to be intellectually subnormal. These include, for example, the sensing of an unfriendly atmosphere in a group of people, or the many acts of judgement which must underlie a stray kitten's manipulation of human emotions as it secures its place in a new household. It is clear that, although human and artificial intelligence compete in many application areas, there are ways in which they have to be seen as having different aptitudes.

II. AI applications related to conflict

Of all the fields of intellectual activity, that of AI is the one in which the change of political environment in the past decade or so has been most dramatic. Until about a decade ago there was a steady stream of exchangees between the Western research centres at MIT, Stanford, and so on, on the one hand, and Moscow and Novosibirsk on the other, with Edinburgh and other West European centres as staging-posts in between. That such freedom was permitted was surprising even then, since it was clear that many AI topics had possible military applications. The potentialities have now been recognized, and in fact overestimated, by governments and the military, with consequent restriction of freedom. The ethics of participation in defence-related work have been debated in the AI literature, notably by Meltzer[2] and Wilks[3] and with a valuable summing-up by Yazdani and Whitby.[4] In the USA the main source of funding for AI research is the military, and undoubtedly much of the overestimation of the military potential stems from extravagant claims in fund-seeking research proposals. The reliability issue (see below) has been discussed by Thompson.[5]

It is convenient to discuss the relevance of AI to conflict under three headings, namely: (*a*) its application in fairly basic ways, as in the navigation of cruise missiles; (*b*) its application where it results in the automation of critical decisions; and (*c*) applications at higher levels of sophistication, including that of international negotiation.

Basic applications of AI

AI techniques can be applied in a basic way in guided missiles, for example, to help steer the missile to its target and determine evasive action when the missile is itself attacked. The possibilities here are endless. Systems to steer a missile to its target once the latter has been sensed are now fairly commonplace and probably not seen as examples of AI. Rather more sophisticated is the means that can be incorporated in missiles for their navigation by recognition of known features of the terrain they are flying over. Evasion tactics can include the release of decoys, and there are various other ways in which intelligent systems might try to deceive their opponents. They might, for example, lull the opponent into false security by pretending, for a time, to be fully occupied with the pursuit of one of his decoys, when in fact the decoy had been recognized as such.

Of course, if wars were fought entirely between armies of robots they would be objectionable only because of the drain on economic resources. Unfortunately it is impossible to believe that a war would terminate without slaughter of human participants at some stage. The conventional view is that a war is not finished until the land forces of one side have physically taken possession of the territory of the other. In modern times it is necessary to consider the further alternative that such occupation is not carried out because the territory of one or both sides has become uninhabitable.

There is little doubt that AI techniques provide, in principle, the means of conducting electronic warfare in highly sophisticated ways. However, there are grave doubts about the reliability of the systems which would be needed, and in fact about the reliability of systems already in existence. Complex information-processing systems of all kinds are normally developed by successive trials, each followed by modifications to eliminate whatever malfunctions have become apparent. It is inevitable that systems for electronic warfare are developed without the possibility of tests under true operational conditions, and their reliability is consequently highly suspect.

There have been numerous attempts to devise rigorous ways of checking program validity. Despite the exercise of a great deal of ingenuity, the available methods are still of purely academic interest in the context of large-scale practical systems. As emphasized by Thompson,[6] failures in complex systems need not result from hardware faults or low-level programming mistakes; the most worrying ones stem from combinations of circumstances not foreseen by the system designers. It is impossible for designers to foresee all possibilities.

Critical applications

Worries about reliability become particularly acute where the AI system is able to initiate a major action such as the launching of a missile. These worries are relevant, for example, in connection with the US Strategic Defense Initiative (SDI), which might involve the use of kinetic energy and beam weapons to intercept missiles and satellites.

A point stressed by Yazdani and Whitby[7] is that AI systems may fail in ways that are specially dangerous because the system continues to have the appearance of working correctly. The danger stems from expectations built up by experience of computers with non-AI programs. In these, malfunctions are usually obvious to a user, and in fact most of them result in a system 'crash' and no coherent output whatsoever. Certain types of failure affecting AI programs can also produce a crash, but there is a greater chance of insidious malfunctions resulting in coherent but invalid output. An expert system with a false entry in its data base or deduction rules might even defend its erroneous conclusions by plausible arguments. Of course, the fact that AI systems are not totally reliable does not necessarily imply that they are less so than human operators; accidental nuclear war could be initiated by a malfunction of either. However, the human embodies evolved characteristics usually denoted by 'common sense' which make him less prone to certain types of error.

Higher-level applications

As mentioned above, AI systems might be responsible for releasing decoys, and perhaps for misleading the enemy by pretending for a time to be fooled by a decoy. The employment of intelligent systems in this kind of activity requires them to be able to model the state of knowledge of other intelligent systems in their environment. Some AI work has been directed at such modelling, for example by Konolige[8] who refers to the 'Problem of the Three Wise Men' as a task demanding it. The task is outlined here in appendix 4A. Konolige describes a computer program capable of acting as the wisest of the three participants, provided the problem is suitably presented to it.

It has even been suggested that AI programs could play a useful part in international negotiations aimed at avoiding armed conflict, but the main argument of the present discussion is that this is extremely unlikely. Before considering this further it is necessary to make more general observations on the nature of AI studies.

III. Well-defined problems

Much of the early work on AI (see Andrew[9]) was restricted to the solution of well-defined problems. These are such that, given something purported to be a solution, there is a systematic way of testing whether it really is one. Theorem-proving in various areas of mathematics has this character. Provided the form a solution could take can be stipulated, it is not difficult, in principle, to write a computer program to generate all possible permutations, and if these are tested in turn a solution will eventually be found if one exists.

As is well known, the method of exhaustive search is impracticable for non-trivial problems because the set of plausible forms of solution is enormous. It is necessary to employ *heuristics*, or rules for directing the search in profitable directions.

Where attention is restricted to well-defined problems there is a sense in which AI work is on safe ground, in that objectives are easily defined and progress can be assessed. However, many real-life problems do not have the convenient property of being well-defined.

IV. Departures from the well-defined problem area

Historically, one of the first contexts in which ill-defined problems had to be considered was that of making programs to play board games such as chess. The problem of choosing a move at some stage in a game would be well-defined if it were possible to consider every single way that the game could continue to its termination, following each alternative move open to the player. This is impracticable, and the choice has to depend on less precise techniques whose only justification is that they seem plausible and have been found to work.

The departure from the well-defined problem area is needed in chess-playing because of the computational complexity of the alternative. There are many other situations in which the problem is ill-defined from the start. Where the requirement is to generate legal or medical advice, for example, the task environment is too complex to allow rigorous analysis, though the value of one decision method can perhaps be compared with that of another by retrospective statistical analysis of the outcomes of following the respective pieces of advice. In recent years the use of *expert systems* (see, for example, Clancey and Shortliffe[10]) has allowed AI to operate in areas in which the problems are clearly ill-defined, because the programs have been made to copy the methods of a human expert.

A special class of ill-defined problem is that of producing aesthetic material, whether as visual art or poetry or music. The merit of a work of art can only ultimately be defined in terms of its effect on a human observer. A human artist therefore has his own built-in means of evaluating his productions, and the corresponding facility incorporated in a program for computer art can only be an approximation to this. Since the artist presumably hopes his work will be appreciated by other observers, he is using himself as a model of their responses.

A great deal of human activity must involve a similar use of the person himself as a model. When two individuals communicate using speech, the form of the utterances is strongly dependent on the speaker's model of the listener and of the latter's view of the world. Non-verbal communication (using winks, nudges or facial expressions) is even more strongly dependent on modelling by the communicator. The use of language has been discussed in these terms by Appelt.[11] The modelling required here has a different flavour to that required for the 'Three Wise Men' problem. The latter requires an impressive application of logic but uses only a small amount of data. The modelling involved in communication has probably less logical depth but can draw on the vast mass of data that constitutes a person's experience of life, or at least on that part of it which is fairly consistent from one individual to another. It is because of the virtual impossibility of providing this mass of data in a computer system that the idea of implementing high-quality mechanical translation between natural languages has been abandoned in the past two decades.

Delicate negotiations between individuals or groups is another task in which people can be expected to act more effectively than machines, because they use themselves as models of the other parties involved. There are features common to negotiation and aesthetic creativity. In both areas, computers can be valuable in ancillary roles, but the ability of a human to use himself as a model of other humans gives human intelligence a powerful advantage. (Some further arguments supporting this contention are introduced below with direct reference to conflict). That people and other animals are so complex that they are difficult to model is confirmed by Tinbergen,[12] whose remarks are as true today as when they were made in 1968: 'in many respects we are still, to ourselves, unknown'.

V. The Prisoner's Dilemma

This name has been given to a simple model of international conflict, discussed at length by Rapoport,[13] as well as by Luce and Raiffa,[14] Genesereth et al.[15] and others noted below. The name stems from the story of two prisoners charged with the same offence, in circumstances where they will both receive a mild punishment if neither confesses, or a relatively severe one if both confess. If one confesses but the other does not, the confessor will be treated leniently in recognition of his co-operation, but the other will be treated more severely than if both had confessed. In the usual version of the story they have to make their decisions without the possibility of communication.

The situation is usually represented in game-theoretic terms by the presentation of a 'pay-off matrix' with rather arbitrary numerical values assigned to the various outcomes. However, it is easy to see, from the informal description, that each prisoner sees his optimal course of action (on a purely selfish basis) as being to confess, despite the fact that the totality of punishment is less if neither follows this course. The situation is often proposed as a model for international tension and armament. Given two countries, each has the choice between arming itself against the other, or remaining unarmed. If each could trust the other and remain unarmed both would benefit, like the two prisoners if neither confesses. The benefits would include (*a*) avoidance of the risk of accidental war, (*b*) a large economic saving, and (*c*) a more pleasant living environment.

Of course, it would be naive to suggest that all the complex aspects of international tension can be represented in such a simple model. Some of it stems, for example, from ancient animosities or feelings about right to possession of territory, and the origins of these are lost in the mists of time. Nevertheless, at least for the current tension between the superpowers, the Prisoner's Dilemma (PD) seems to be a fairly good representation of the state of affairs.

While it is true that every sane person wants peace, there are many parties who find it advantageous to play a game of *brinkmanship*. Among these are vested interests in the armed forces and defence-related industries, and also national political leaders, who find they are applauded by their electorates for 'taking a firm stand'. There is also a tendency for internal conflicts in a country to be set aside when there seems to be a common enemy (Tinbergen[16] discusses this as a phenomenon seen in all group-territorial species), and so brinkmanship makes each country easier to govern. The situation is further aggravated by the stereotyped images of people of the other nationality sustained by writers of film scripts and other fiction—a harmonious international situation does not make for exciting theatre!

These various influences do not destroy the relevance of the Prisoner's Dilemma model, but they influence the subjective estimates of appropriate entries in the pay-off matrix in a way that aggravates the situation.

VI. Linguistic analysis

Much effort in AI has been devoted to the automatic interpretation of natural language. It is customary to distinguish the deep structure of a sentence from its surface structure—see for example McKeown.[17] However, this work is aimed at unravelling the speaker's conscious intentions, not at detecting deep motives in the way that Siegfried, in Wagner's opera, was able to do after he had tasted the dragon's blood. National leaders are fond of making ringing declarations such as 'We will negotiate from a position of strength', or 'It is our strength that has brought them to the conference table'. Unfortunately, something more than linguistic analysis is needed to show that these have a deeper meaning: 'We are guilty in accelerating the arms race'.

VII. Getting around the Prisoner's Dilemma

Given that the PD model is valid, the only hope of disarmament is a degree of trust between the participants, and it is difficult to see how this can be achieved other than by negotiation between human participants. As argued earlier, humans have a fundamental advantage over machines in the context of negotiation, since they are able to use themselves as models of their opposite numbers.

There is a connection here with the observation made by Tinbergen,[18] and many others, that war activities become increasingly difficult the greater the communication between the opponents. For most people it is much easier to kill others with the help of modern technology, by dropping bombs or releasing missiles, than it is to stab them or shoot them when they can actually be seen. Man has evolved certain characteristics of altruism which come into play when the opponent is seen to have human characteristics. Similarly, the means of establishing trust have been evolved, but only become effective when the opposite number is seen to be human. It it therefore unlikely that artificial intelligence techniques will play any central role in negotiation.

The question of how altruistic behaviour could arise in evolution has been much discussed by biologists, notably Haldane,[19] Hamilton[20] and more recently Ozinga.[21] There are many situations in life having the characteristics of the Prisoner's Dilemma, and communal activities are only possible because individuals do not respond to these situations with complete selfishness. A recent study by Axelrod,[22] discussed also by Hofstadter,[23] suggests how altruistic behaviour could have evolved.

Axelrod simulated a tournament in which a large number of interacting entities met each other repeatedly in a PD-like situation. The entities were able to recognize one another, so their actions on meeting another individual could be dependent on their previous experience with that individual. Axelrod obtained from a number of experts (14, to be exact), suggestions for a policy to be followed by entities taking part. The policies were defined by sections of a computer program in the Basic language. He added a fifteenth policy, which was that of random and equiprobable choices between the two forms of response. He then arranged the tournament in which each of the 15 policies was made to interact with each of the others, as well as with a copy of itself, making 200 simulated meetings in each combination.

In each meeting between two entities the pay-off matrix associated with the PD model was used to determine the respective pay-offs. The total pay-off was formed for all the meetings engaged in by each of the 15 policies. This can be seen as a simulation of an evolutionary situation in which individual animals acted with different degrees of altruism. It is reasonable to regard the total pay-off as an indication of the evolutionary advantage of a policy.

An interesting and surprising outcome was that the policies which were most successful were not the most selfish ones. That is to say, they were not those which strongly favoured the distrustful choice corresponding to a prisoner deciding to confess, or a nation deciding to arm. A policy was termed 'nice' if it could never be the first to act in this way in its interactions with another particular policy—in other words, it would never behave distrustfully unless provoked. In the tournament arranged by Axelrod, 'nice' policies achieved high scores.

The experiment shows that the optimal policy for a multi-stage PD process is not the same as the selfish one which is optimal for a single stage. The evolution of altruism becomes understandable. Unfortunately the power of modern weapons is such that the superpowers cannot afford to engage in a multi-stage PD process over the disarmament issue. The findings therefore reinforce the view that humans must get together for negotiations, so that the altruism built into them by their evolutionary background can favour 'nice' policies. In this way the decision in the current international situation may be treated, not as a single-stage PD process but as one further move in a lengthy multi-stage one. Despite the undoubted achievements of work on artificial intelligence, international negotiations must depend on peculiarly human qualities.

Notes and references

[1] Forsyth, R. and Naylor, C., *The Hitch-Hiker's Guide to Artificial Intelligence* (Chapman and Hall: London, 1985), preface.

[2] Meltzer, B., 'AI and the military', *AISB—Quarterly Newsletter of the Society for the Study of Artificial Intelligence & Simulation of Behaviour*, no. 52 (1985), pp. 24–26.

[3] Wilks, Y., 'AI and the military again', *AISB*, no. 53/54 (Summer 1985), pp. 23–24.

[4] Yazdani, M. and Whitby, B., 'Accidental nuclear war: the contribution of artificial intelligence', Working Paper no. W145, Computer Science Department, University of Exeter, England, 1986. (A shortened version appears in *AISB*, no. 58 (Summer 1986), pp. 6–7).

[5] Thompson, H., 'There will always be another moonrise', *AISB*, no. 53/54 (Summer 1985), pp. 21–23. (The title is a reference to a false warning of attack produced by radar signals reflected from the moon, illustrating that it is impossible for system designers to foresee all possible circumstances. The same point is made strongly in chapter 7).

⁶ See note 5.
⁷ See note 4.
⁸ Konolige, K., 'A first-order formalisation of knowledge and action for a multi-agent planning system', in Hayes, J. E., Michie, D. and Pao, Y.-H. (eds), *Machine Intelligence 10* (Ellis Horwood: Chichester, 1982), pp. 41–72.
⁹ Andrew, A. M., *Artificial Intelligence* (Abacus: Tunbridge Wells, 1983).
¹⁰ Clancey, W. J. and Shortliffe, E. H. (eds), *Readings in Medical Artificial Intelligence* (Addison-Wesley: Reading, MA, 1984).
¹¹ Appelt, D. E., *Planning English Sentences* (Cambridge University Press: Cambridge, 1985).
¹² Tinbergen, N., 'On war and peace in animals and man', in: McGill, T. E. (ed.), *Readings in Animal Behaviour* (Holt, Rinehart and Winston: New York, 1977), chapter 53, pp. 452–67.
¹³ Rapoport, A., *Fights, Games and Debates* (University of Michigan Press: Ann Arbor, 1961); Rapoport, A., 'Prisoner's Dilemma', in Rapoport, A. (ed.), *Strategy and Conscience* (Shocken Books: New York, 1969), chapter 6, pp. 48–57.
¹⁴ Luce, R. D. and Raiffa, H., *Games and Decisions* (Wiley: New York, 1957), pp. 94–102.
¹⁵ Genesereth, M. R., Ginsberg, M. L. and Rosenschein, J. S., *Cooperation without Communication*, Report No. HPP-84-36, Stanford University, California, Sept. 1984.
¹⁶ See note 12.
¹⁷ McKeown, K. R., *Text Generation* (Cambridge University Press: Cambridge, 1985), p. 84.
¹⁸ See note 12.
¹⁹ Haldane, J. B. S., 'Population genetics', *New Biology*, no. 18 (1955), pp. 34–51.
²⁰ Hamilton, W. D., 'The genetical evolution of social behaviour', *Journal of Theoretical Biology*, no. 7 (1964), pp. 1–16 and 17–52.
²¹ Ozinga, J. R., 'Genetic altruism', *Humanist*, vol. 41 (1981), pp. 22–27.
²² Axelrod, R., *The Evolution of Cooperation* (Basic Books: New York, 1984).
²³ Hofstadter, D. R., 'Metamagical themas: computer tournaments of the Prisoner's Dilemma suggest how cooperation evolves', *Scientific American*, vol. 248, no. 5 (May 1983), pp. 14–21; Hofstadter, D. R., 'Metamagical themas: the calculus of cooperation is tested through a lottery', *Scientific American*, vol. 248, no. 6 (June 1983), pp. 14–19.

Appendix 4A. The problem of the three wise men

There was once a king who had three wise men, and wanted to decide which was the wisest. On the forehead of each he made a mark, either black or white, such that each could see the marks of the other two but not his own. He also told them (and each knew that the others had been told) that there was at least one white mark. In fact all three marks were white. The wisest of the three would be the first to declare the colour of his own mark. There was no communication between the three men, except that each would know if another solved the problem first.

After a time, one of the men gave the right answer. How was this possible?

If the successful participant is termed A, he was able to reason as follows. 'If my spot is black, one of my colleagues, say B, would be able to say that if his (B's) spot was also black then C would know that his (C's) was white. Hence, B would be able to infer from the silence of C that his (B's) spot must be white. But B is silent, so the initial premise, that my (A's) spot is black, must be erroneous. Therefore my spot is white.

Part III. Military and strategic implications

Chapter 5. The Strategic Computing Program

S. INGVAR ÅKERSTEN

I. Background

The Strategic Computing Program (SCP) was announced in Washington, DC on 7 November 1983 by the US Defense Advanced Research Projects Agency (DARPA). The substance of the plan was summarized as follows:

- Program Objectives
  —Provide a broad base of machine intelligence technology for application to critical defense problems
  —Create a strong industrial capability to support national security requirements
- Military Challenge
  —Demonstrate machine intelligence technology on critical future military systems such as autonomous vehicles, expert associates and battle management
  —Develop capabilities to allow intelligent machines to survive in hostile environments such as the battlefield and in space
  —Improve our ability to respond to unpredictable military situations quickly
- Technological Challenge
  —Develop a new generation of computers that can SEE, HEAR, TALK, PLAN and REASON
  —Leverage advances in artificial intelligence and micro-electronics coupled with computer architectures and software
  —Realize the power in multiple computers working together
- Program Management and Funding
  —Management by the Defense Advanced Research Projects Agency (DARPA)
  —In close cooperation with the military services and Defense agencies
  —Program initiated in FY84—funding set at $50M in FY84, $95M in FY85[1]

Speculation about the origin and *raison d'être* of the programme took two main directions. On the one hand some observers presumed this initiative to be an answer to Japan's Fifth Generation Computing Project, announced about two years earlier. It was believed that the two were sufficiently similar in spirit to justify such a presumption. On the other hand some observers believed that the SCP was set up to serve as the development agent for the Strategic Defense Initiative (SDI) that President Reagan had announced in his 'Star Wars' speech on 23 March 1983.

Several DARPA representatives, including the director and also the SCP manager, have maintained that the way in which these three initiatives seem to coincide in subject matter and time is essentially just that—coincidental. To some extent these assurances were borne out about a year later when SDI was made an independent programme. At that time both the responsibility and the funds for SDI were cut from the DARPA Information Processing Techniques Office (IPTO) without seemingly affecting the direction and pace of the SCP.

II. DARPA

DARPA started out as the Advanced Research Projects Agency in February 1958. Together with the National Aeronautics and Space Administration (NASA) and the Department of Defense (DoD) Office of Research and Engineering, it was founded in response to the early accomplishments of the Soviet space programme, which had 'startled' the world with the launch of Sputnik I on 4 October and Sputnik II on 3 November 1957. Its founding was, however, preceded by the launch of the first US satellite, Explorer I, on 31 January 1958 which, together with Explorer III launched on 26 March 1958, provided the measurements which led to the discovery of the van Allen radiation belts.

As opposed to many other agencies, including NASA, DARPA has not itself been engaged in research and development work. It has—through programme managers—designed, funded and overseen concerted research and development (R&D) efforts in academia, non-profit 'think-tank' organizations and industry.

Although DARPA is a defence body, with partly military personnel and funds and bearing the motto 'Ensuring leadership in military technology', it has over the years been engaged mainly in the establishment and long-term development of basic (generic) technologies rather than of specific military applications work. Unquestionably most of this newly established and evaluated knowledge has been 'dual use' in character, that is, it has had potential applicability to civilian as well as military products. Furthermore, there have been no obvious formal limitations on the commercial use of the results obtained from DARPA programmes—on the contrary. It seems to have been common practice for commercial contractors to quickly incorporate results from DARPA-funded projects into their own proprietary products. Some results may in fact have found predominantly civilian applications rather than military ones.

Since its establishment in 1962 DARPA's IPTO has been considered by many to be the dominant force in the development of computer science to its present extent and direction. At that time computing was at a very tender age. The first electronic calculator ENIAC, from 1946, had in 1948 been partially modified in accordance with the computer architecture ideas of John von Neumann. The first computer originally constructed entirely according to von

Neumann's principles was produced in England in 1949 and the first in the USA, built under the direction of von Neumann himself, was initially operational in 1951 and finally dedicated in 1952. Most of the computer techniques in common use today were conceived and initially tested during the 1950s but much R&D work remained to transform these ideas into a viable technology. DARPA came to play an important role in this process. It became a major force through being willing and able to invest in high-risk enterprises while other funding agencies, or industry, were unwilling or unable to invest for various reasons. Some programmes have resulted in entirely new techniques or ways of doing things but, at the same time, the probability for spectacular failures has been rather large. As a result of its high-risk policy DARPA has also had its fair share of such failures to match its impressive accomplishments.

Among the fields in which DARPA-funded research has laid the groundwork are interactive computing, timesharing, and computer networking, and it has also contributed to fields such as artificial intelligence (AI), natural-language programming, image/speech interpretation and other generic 'fifth-generation' type technologies.

During it's first two decades DARPA was able to concentrate its efforts on the pure and applied research and development phases of the areas of interest. Once an idea had been shown to be realizable during these phases, the interested parties—military services or industry—took over the advanced technical implementation. Since the end of the 1970s DARPA has seemingly had to take over this latter role more and more. This seems to be so to an even greater extent in the current SCP case. The plan also calls explicitly for DARPA management of the technology transfer of the results of this programme into the military services. In part this will take the form of cost-sharing with the Services in the development of military applications. DARPA does however stress that technology transfer to industry is equally important and that, to this end, it 'will make full use of regulations for Government procurement involving protection of proprietary information and trade secrets, patent rights, and licensing and royalty arrangements'.[2]

Perhaps the gradual change in the DARPA role was an inevitable consequence of changing Federal or Services appropriations priorities. Perhaps the radically new DARPA role in the SCP is a catalytic effect of the Japanese Fifth Generation Project which may have opened people's eyes to the fact that more systematic use can be made of the results of DARPA-funded R&D work than had hitherto been customary. It seemed an obvious conclusion that DARPA itself would be the US Agency best suited to assume this new role. DARPA would probably sooner or later have assumed the role in any case—perhaps even in response to demands from the military services who would themselves have found it increasingly difficult to uphold such a role as the complexity of the new DARPA-sponsored findings grew. Whatever its cause, the new role and some of the programme's contents, such as 'autonomous techniques', have been, and remain, a matter of controversy among some scientists and other interested, concerned or otherwise affected parties.

III. The strategic computing plan

A comprehensive overview of the proposed programme is given in the *'Strategic Computing'* document dated 28 October 1983,[3] some highlights of which are briefly summarized below.

The rationale and potential achievements of the programme are outlined in the introduction. It argues that:

As a result of a series of advances in artificial intelligence, computer science, and microelectronics, we stand at the threshold of a new generation of computing technology having unprecedented capabilities. The United States stands to profit greatly both in national security and economic strength by its determination and ability to exploit this new technology.

Computing technology already plays an essential role in defense technologies such as guided missiles and munitions, avionics, and C^3I. If the new generation technology evolves as we now expect, there will be unique new opportunites for military applications of computing. For example, instead of fielding simple guided missiles or remotely piloted vehicles, we might launch completely autonomous land, sea, and air vehicles capable of complex, far-ranging reconnaissance and attack missions. The possibilities are quite startling, and suggest that new generation computing could fundamentally change the nature of future conflicts.

In contrast with previous computers, the new generation will exhibit human-like, 'intelligent' capabilities for planning and reasoning. The computers will also have capabilities that enable direct, natural interactions with their users and their environments as, for example, through vision and speech.

Using this new technology, machines will perform complex tasks with little human intervention, or even with complete autonomy. Our citizens will have machines that are 'capable associates', which can greatly augment each person's ability to perform tasks that require specialized expertise. Our leaders will employ intelligent computers as active assistants in the management of complex enterprises. As a result the attention of human beings will increasingly be available to define objectives and to render judgments on the compelling aspects of the moment.

A very broad base of existing technology and recent scientific advances must be jointly leveraged in a planned and sequenced manner to create this new intelligent computer technology. Scientists from many disciplines, scattered throughout the universities, industry, and government must collaborate in new ways, using new tools and infrastructure, in an enterprise of great scope. Adaptive methods of planning must be applied to enhance the process of discovery. Events must be skillfully orchestrated if we are to seize this opportunity and move toward timely success.

In response to these challenges and opportunities, the Defense Advanced Research Projects Agency (DARPA) proposes to initiate an important new program in Strategic Computing. To carry out this program, DARPA will fund and coordinate research in industrial, university, and government facilities, and will work with the Military Services and Defense Agencies to insure successful transfer of the results.

The overall goal of the program is to create a new generation of machine intelligence technology having unprecedented capabilities and to demonstrate applications of this technology to solving critical problems in Defense. Although the achievements of the program applications' objectives will significantly improve the nation's military capabilities, the impact of nonmilitary spin-offs on the national économy should not be underestimated.[4]

This sketchy vision or general policy statement lays the foundation for the ensuing detailed descriptions of military challenges, technical opportunities, goals, methods, activities and plans.

As a result of 'growing complexity of forces and rising level of threats' ever more advanced 'computers are being increasingly employed to support United States military forces'. An obstacle in this process is 'the rapidly decreasing predictability of military situations, which makes computers with inflexible logic of limited value'. Therefore 'access to adaptive technology is considered important to defense'. Against the present situation in these respects it was deemed obvious that 'intelligent military systems demand new computer technology'.

'The trend in all areas toward faster-moving warfare severely stresses the whole staff function. Greater uncertainty in the military environment forces consideration of more options'. However, 'current computers provide only limited assistance to such decision making because they have limited ability to respond to unpredictable situations and to interact intelligently with large human staffs'.

At present, however, the situation is such that 'many isolated pieces of the required technology are already being developed. The challenge is to exploit these beginnings, make new efforts to develop the full set of required technologies, and integrate components of the emerging new technology in order to create revolutionary defense capabilities'.

Examples were given of some of the important advances which facilitate the creation of a new generation of computing technology:

KEY AREAS OF ADVANCES THAT CAN BE LEVERAGED TO PRODUCE HIGH-PERFORMANCE MACHINE INTELLIGENCE

- Expert Systems: Codifying and mechanizing practical knowledge, common sense, and expert knowledge.
- Advances in Artificial Intelligence: Mechanization of speech recognition, vision, and natural language understanding.
- System Development Environments: Methods for simplifying and speeding system prototyping and experimental refinement.
- New Theoretical Insights in Computer Science.
- Computer Architecture: Methods for exploiting concurrency in parallel systems.
- Microsystem Design Methods and Tools.
- Microelectronic Fabrication Technology.

Properly combined all these recent advances now enable us to move toward a completely new generation of machine intelligence technology.

In addition to the planned military applications the value of future commercial products made available by development of the new generation technology will be enormous.[5]

The logical structure of the SCP and its goals are visualized in the *Strategic Computing* document as shown in figure 5.1.

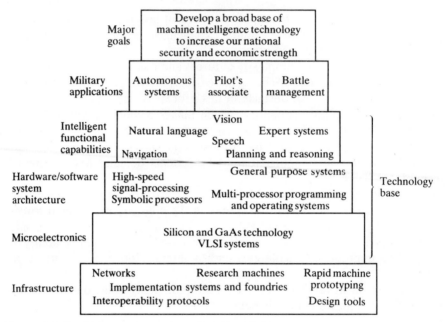

Figure 5.1. Logical structure and goals of the SCP
Source: Conway, L. (ed.), *Strategic Computing—New-Generation Computing Technology: A Strategic Plan for its Development and Application to Critical Problems in Defense* (US Defense Advanced Research Projects Agency: Arlington, VA, 28 Oct. 1983), p. 11.

A relatively broad base of disparate computing capabilities which were ripe for more systematic exploitation was the starting point for the SCP. The method of evolving these capabilities into a mature new generation of infrastructure base technology was to create an integrated planning framework combining the setting of realistic near-term technology development goals with the proper selection of application demonstrations. This means that the first phase of the SCP stood on firm ground and that it was possible to make rather well-founded estimates of the near-term accomplishments.

The three military applications initially selected were an Autonomous Land Vehicle, a Pilot's Associate and a Battle Management System for a Carrier Battle Group. In the words of the *Strategic Computing* document:

These specific examples are included in the Strategic Computing program based on a weighted consideration of the following factors:
- The application must effectively employ the new technology to provide a major increase in defense capability in light of realistic scenarios of combat situations that might occur at the future time when the new systems can be procured and deployed.

- The application must provide an effective 'pull' on the new generation technology. It must demand an aggressive but feasible level of functional capability from one or more of the intelligent functions at appropriate points in the timeline.
- Development of the application must lead to new engineering know-how in artificial intelligence software areas, such as planning and reasoning, learning, navigation, knowledge base management, and so on.
- The application must test the efficacy of the new technology at a realistic quantitative scale of performance demands. In this way we seek to ensure against unexpected quantitative changes in system performance as a result of scaling up from models and laboratory experiments to real systems.
- The application must provide an effective experimental 'test-bed' for evolving and demonstrating the function(s). Stability over time, access, and visibility are thus important factors.
- The application must effectively leverage program resources. Thus an important factor is the extent to which an existing military program provides a base of capital resources and experienced personnel into which the new generation technology can be experimentally introduced (versus this program having to provide such non-computing resources).
- It is important to choose a mix of applications that are jointly supportive of and involve all three Services, and which are appropriately executed through each Service. Only in this way can we develop the base for extension of this technology into a wide range of military systems.
- Finally, an important selection factor is the potential provided by the specific application for effecting the transfer into the services of the new machine intelligence technology.

The choices have been initially made on this basis, but it is recognized that further planning and the evolving technology development may lead to a change in the choice of specific application demonstrations. It might, for example, prove preferable to pursue an autonomous underwater vehicle rather than a land vehicle, and a battle management system for land combat might prove more appropriate than that for the Naval application. A panel of the Defense Science Board has been convened to make recommendations on how best to exploit machine-intelligence technology within the DoD, and that panel will be providing information and advice for this program. Consequently, we anticipate that some of the specifics may change over time, within the framework described.[6]

The goal-setting process may be illustrated by referring to one of these application experiments and demonstrations, say the autonomous land vehicle. One could imagine a reconnaissance vehicle that could autonomously navigate a given maximum distance. It would have to:

be capable of planning an initial route from digital terrain data, updating its plan based on information derived from its sensors, resolving ambiguities between sensed and pre-stored terrain data, and incorporating landmark prediction and identification as navigation means. Using advanced image understanding technology, the reconnaissance payload would perform image segmentation and other basic scene processing upon arrival in a designated area, identify target objects, and report its findings and interpretations. To develop an autonomous land vehicle with the capabilities described requires an expert system for navigation and a vision system.[7]

Assuming that these functions are carried out in real time while the vehicle is moving cross-country at high speed, and considering the complexity of such tasks, certain quantitative requirements have to be fulfilled. These may relate to the volume of information stored, the rate of inference making, the machine processing rate, weight, volume, power dissipation, and so on. Quite a few such quantitative assessments are to be found in the *Strategic Computing* document; their level of detail is beyond the needs of this presentation. They do however provide an excellent means for the programme management to follow up goal fulfilment at several points in time.

In terms of the technology base certain functional objectives were put down as sub-goals to be accomplished in different fiscal years during the course of the proposed programme, such as:

Functional Objectives for Vision Subsystems
FY86 Model and Recognize Simple Terrain with Crude Objects
FY88 Recognize and Match Landmarks with Maps in Simple Terrain
FY90 Recognize and Match Landmarks and Obstacles in Complex Terrain Using Rich Object Descriptions
FY92 Perform Reconnaissance in a Dynamically Changing Environment

Functional Objectives for Speech Subsystems
FY86 Recognition of Words from a 100-Word Vocabulary for a Given Speaker under Severe Noise and Moderate Stress Conditions
FY88 Recognition of Sentences from a 1,000-Word Vocabulary with Moderate Grammatical Constraints in a Speaker Adaptive Mode under Low Noise and Stress Conditions
FY89 Recognition of Connected Speech, Independent of Speakers from a 200-Word Vocabulary with Strict Grammatical Constraints under Severe Noise and High Stress Conditions
FY92 Recognition of Sentences, Independent of Speakers, from a 10,000-Word Vocabulary with Natural Grammar under Moderate Noise and Low Stress Conditions

Functional Objectives for Natural Language Subsystems
FY86 Natural-language interfaces with some understanding of commands, data inputs, and queries (e.g., interface to a database and a threat-assessment expert system)
FY88 Domain-specific text understanding (e.g., understand paragraph length intelligence material relating to air threat)
FY90 Interactive planning assistant which carries on task-oriented conversation with the user
FY93 Interactive, multi-user acquisition, analysis, and explanation system which provides planning support and substantive understanding of streams of textual information.[8]

These functional objectives can of course also serve as a basis for goal fulfilment follow-ups.

The SCP was intended to be a 10-year programme running from 1984 until

Table 5.1. Strategic computing cost summary
Figures are in US $m.

	FY84	FY85	FY86	FY87[a]	FY88[a]
Total military applications	6	15	27	TBD	TBD
Total technology base	26	50	83	TBD	TBD
Total infrastructure	16	27	36	TBD	TBD
Total programme support	2	3	4	TBD	TBD
Total	50	95	150	TBD	TBD

[a] Out-year funding levels to be determined (TBD) by programme progress.
Source: Strategic Computing—New-Generation Computing Techology: A Strategic Plan for its Development and Application to Critical Problems in Defense (US Defense Advanced Research Projects Agency: Arlington, VA, 28 Oct. 1983), p. 66.

the end of 1993. Initially it was only funded for the first five years, as shown in table 5.1. Total programme costs for the first five-year period were estimated at approximately US $600 million. Nothing was mentioned about the rate of funding during the last five years other than an estimated $150 million per year for a several-year period around the peak of the programme.

The SCP management is handled by the DARPA IPTO, sharing the responsibility with other offices, for example in the areas of microelectronics and applications.

Because of its intended national importance the programme is supposed to be co-ordinated with organizations involved in related technologies. By spring 1983 an agreement for exchange of information had been reached between Department of Commerce (DOC), DoD, Department of Energy (DOE), NASA, National Science Foundation (NSF) and Office of Science and Technology of the President (OSTP) representatives. The first information-exchange meeting was held in June 1983 and it seems that the material presented then must have been akin to that of a previous DARPA announcement that it was initiating a Strategic Computing and Survivability Program.[9] A panel of the Defense Science Board (DSB) has also been convened as an advisory body alongside the defence agencies such as the Under Secretary of Defense for Research and Engineering (USDRE) and the individual military services.

IV. Initial SCP achievements

A review of the SCP half-way through the first five-year period gave a comprehensive and very positive picture of the accomplishments of the programme up to that point.[10] Until then the SCP was probably running more or less as could have been predicted at the outset. The reason for this is that many of the activities within the SCP seem to have been immediate successors or extensions of previous DARPA-funded projects or others within the 'DARPA realm'. A total of 76 SCP projects were reported in the review, as shown in table 5.2.

Table 5.2. Current strategic computing projects

Activity	Project
Autonomous Land Vehicle	Integration Terrain Data Base ALV Route Planning Research Optical Avoidance & Path Planning Telepresence System
Navy Battle Management	Force Requirements Expert System (FRESH) Spatial Data Management System (SDMS/VIEW) Combat Action Team (CAT) Fleet Command Center Battle Management System Combat Action Team (CAT) System
Pilot's Associate	Pilot's Associate Development Phase I (Dual Award)
AirLand Battle Management	System Technology Definition Soldier–Machine Interface
Multiprocessor System Architectures	Tree Machines Software Workbench Programmable Systolic Array Ada Compiler System Synchronous Multiprocessor Architecture High Performance Multiprocessor VLSI Design Common LISP Framework Data Flow Emulation Facility Massive Memory Machine Connection Machine
Natural Language Understanding	New Generation System Language Understanding Language Generation Context of Discourse Language Generation Language Understanding Japanese Translation Processing Written Discourse
Expert System Technology	Expert System Knowledge Reasoning with Uncertainty Evolutionary New Generation System Tool Reasoning with Endorsement Revolutionary New Generation System Tool Explanation for Expert Systems Multiprocessor Architecture for Expert Systems
Speech Understanding	Continuous Speech Understanding Real Time Speech Continuous Speech Understanding Auditory Modelling Acoustic Phonetic-Based Speech Speech Data Base Acoustic Phonetics Tools for Speech Analysis Speech Data Base Robust Speech Recognition Speech Co-Articulation Speaker Independence

Table 5.2. cont.

Activity	Project
Computer Vision	Vision Based Navigation
	Parallel Algorithms
	Terrain Following
	Dynamic Image Interpretation
	Target Motion & Tracking
	Reasoning, Scene Analysis
	Parallel Algorithms
	Spatial Representation/Modelling
Infrastructure	Adaptive Wafer Scale Integration
	MOSIS Service
	Compact LISP Machine
	Butterfly Multiprocessor
	Computer Benchmarking
	GaAs Process Evaluation & Design Rules
Microelectronics	Super Lattice
	Fiber Optic Coupler
	Hybrid Optical Interconnects
	Time-Division Optical Multiplexing
	GaAs Multifunction Arrays
	VLSI Components
	Second Pilot Line
	Silicon GaAs for Optical Interconnection
Program Management	Technical & Management Support

Source: Based on Judge, J. F., 'SCP gets high marks at midterm', *Defense Electronics*, May 1986, p. 66.

One-quarter of the way into the total programme the number of application experiments and demonstrations numbered five, the Army AirLand Battle Management and Strategic Computing for Image Understanding Program (SCIUP)—the latter directed towards radar/optical imagery analysis—having been added to the three initially selected. Details available permit comparison with the functional objectives quoted above.

In the new generation of vision systems, vision algorithms have been demonstrated to run 400 times faster than any previous system through the use of the new parallel processing architectures. . . . For the first time, a 250-word speaker-independent continuous speech recognition with a large vocabulary was demonstrated. The vocabulary will expand to 1000-word recognition by 1987. . . . The new generation natural-language system being developed is the first system to flexibly couple an English front end to both a data base and an expert system. The system exceeds any other existing computer-based natural language ability and understands 2000 word types and 700 domain concepts. By 1987, DARPA expects these to grow to 7000 words and 2500 concepts.[11]

V. The Second Strategic Computing Program plan

Three years after its inception it appears that the SCP has entered a new phase. The Second Strategic Computing Program (SSCP) plan is a loose-leaf document for internal use, in contrast to the agency's original plan for the programme. It is expected that this document will be revised every six months.[12] The SSCP is reported to be budgeted at about $150 million annually which is in accordance with the original intentions.

'We have lived off research advances that preceded DARPA's Strategic Computing initiative. We are now up against the stops on whether we can meet milestones for the next two years', the special assistant to the Director of DARPA for Strategic Computing, Dr Kelly, is quoted to have said.[13] Still another application experiment and demonstration, the sixth, has been added—an Autonomous Smart Weapons Program. All of these application projects are included pursuant to the tenet that the SCP is 'structured around the imperative of technology transfer'. These remarks are quoted from Dr Kelly as he addressed participants at the opening of the second annual conference on aerospace applications of artificial intelligence. On the same occasion he offered information shedding some light on the further accomplishments of the SCP in a rather general fashion, stating, for example, that the Autonomous Land Vehicle 'keeps on meeting its milestones', and that 'we now have the capability to type in essentially natural language queries, and plan to add a speech interface to the natural language capability'.

Progress has also been reported in a project labelled Revolutionary New Generation System Tool. It is being carried out by Teknowledge whose principal scientist is quoted to have said: 'Abe (*A better environment*) is an architecture for building systems intelligently—using what we know about AI to build systems that are perhaps 20% AI and 80% conventional software through what we call knowledge-processing functions'. DARPA's project manager for Abe is quoted to have said that 'the architecture's greatest strength is that other systems can be used within it—Abe provides a common interface to the entire computing world'.[14]

A recent journalistic assessment of the SCP gives an excellent summary of the industrial and academic contractors in the programme.[15] Interestingly the budget for fiscal year 1987 is reported to be only $104 million as opposed to the originally planned and recently reported annual $150 million during peak funding years.

VI. Afterthought

Among the non-AI professionals there has been much concern about the term artificial intelligence, its definition, connotations and usage. It is perhaps worth pointing out that the composite term 'intellectoid' can be formed from: oid—having the form or appearance of [something resembling a (specified) object or having a (specified) quality], and intellect—the power or faculty of knowing as distinguished from the power to feel and to will.[16]

When combined and used in this sense one could conceive and formulate the following, tentative definition: intellectoid systems* are systems having the appearance of the power of knowing.

Notes and references

[1] US DARPA, *Point Paper—DARPA Strategic Computing Program*, a fact sheet handout accompanying: *Strategic Computing Plan Announced; Revolutionary Advances in Machine Intelligence Technology to Meet Critical Defense Needs* (US Defense Research Advanced Projects Agency: Washington, DC, 1983).

[2] Conway, L. (ed.), *Strategic Computing—New-Generation Computing Technology: A Strategic Plan for its Development and Application to Critical Problems in Defense* (US Defense Advanced Research Projects Agency: Arlington, VA, 28 Oct. 1983), Executive Summary, p. viii.

[3] Note 2.

[4] Note 2, pp. 1–2.

[5] Note 2, pp. 6–9.

[6] Note 2, pp. 19–20.

[7] Note 2, pp. 21–22.

[8] Note 2, pp. 34, 36 and 38.

[9] Cooper, R. S. and Kahn, R. E., 'Strategic computing and survivability', *Signal*, June 1983, pp. 25–28.

[10] See Judge, J. F., 'SCP gets high marks at midterm', *Defense Electronics*, May 1986, pp. 65–77.

[11] Note 10, p. 65.

[12] Stein, K. J., 'New Strategic Computing Plan details programs, fiscal data', *Aviation Week & Space Technology*, 15 Dec 1986, pp. 95–99.

[13] Note 12, p. 95.

[14] Schneiderman, R., 'A new way to move AI into the mainstream', *Electronics*, 22 Jan. 1987, pp. 90–92.

[15] Corcoran, E., 'Strategic Computing: a status report', *IEEE Spectrum*, Apr. 1987, pp. 50–54.

[16] Both definitions are taken from *Webster's Third New International Dictionary* (Merriam: Springfield, MA, 1961), pp. 1174 and 1568, respectively.

*Components, computers, methods, modules, techniques, . . .

Chapter 6. Artificial intelligence and the automated tactical battlefield

RANDOLPH NIKUTTA

I. Introduction

Developments in certain key basic technologies are having a profound effect on military technology and are revolutionizing the face of modern tactical warfare. These technologies are revolutionary in their effect because they automate the battlefield and further radically dehumanize the course of war; they are already finding applications in military doctrines, and to a certain degree in the force posture, of the USA and NATO.

In section II the history of the automated battlefield and the present extent of automation in the Western Alliance's armed forces is briefly described. It is noted how military interest in artificial intelligence (AI) emerged in the early 1980s and how it came to be believed that it offered solutions to a broad range of battlefield problems. Examples of possible future applications of AI on the tactical battlefield are outlined and some problems in the application of AI technology to the military domain are addressed. In the light of new doctrinal concepts and possible applications of AI technology, section III examines the command and control requirements for tactical nuclear weapons on the automated battlefield. Some of the potential problems are also analysed. Section IV deals with the escalation risks associated with new Western warfighting doctrines for the automated battlefield.

II. Artificial intelligence and automation on the tactical battlefield

The automated battlefield

Military research and development (R&D) for tactical warfare focuses on the application of advanced technologies—for example microelectronics, microelectronic image-sensing techniques, computer and software technology—to the automated battlefield. This R&D effort draws on the efforts undertaken in the past 15 years to automate ground, air and sea combat, and upon the recent rapid progress in the field of microelectronics.

The evolution of the automated battlefield

The present stage in the evolution of the automated battlefield began during the Viet Nam War when electronic weaponry was developed and tested for the first time. Although the war in Asia turned out to be a major political débâcle for the USA, its one great 'success' was the extraordinary progress in military technology: the development of complex, long-range sensor-technology for surveillance, remotely piloted vehicles, TV- and laser-guided 'smart' bombs. The computer increasingly became a major element of all sophisticated weapon systems and military operations on the tactical battlefield.[1] In 1969, in the light of the promise of such technology, General Westmoreland, Chief of Staff of the US Army, outlined for the first time in public the development of the automated battlefield as an official military goal to be realized within the next 10 years. He described the automated battlefield as one that would be characterized by the following features:

On the battlefield of the future, enemy forces will be located, tracked, and targeted almost instantaneously through the use of data-links, computer assisted intelligence evaluation, and automated fire control. With first round kill probabilities approaching certainty, and with surveillance devices that can continually track the enemy, the need for large forces to fix the opposition physically will be less important . . . the future portends a more automated battlefield.[2]

Westmoreland's vision of a fully 'electronicized' army is beginning to become a reality in the mid-1980s. Microelectronics and their applications in computer technology are playing an increasingly dominant role in today's armed forces for the tactical battlefield.

Technology and doctrine

The development of the automated battlefield throws interesting light on the relationship between technology and doctrine. It is the opinion of the author that no simple relationship exists, but that military doctrine is shaped by a variety of factors. These include new threat perception analyses, changing political and economic conditions, and inter-service rivalries. Nevertheless, as this paper demonstrates, in the case of the automated battlefield, technology has to a large degree shaped military doctrine.

New operational concepts for conducting air and ground warfare were developed by NATO in response to the introduction of new military equipment and to the potential of advanced, so-called emerging, technologies (ET). These new concepts, such as the AirLand Battle (ALB) doctrine or the Follow-On-Forces-Attack (FOFA) concept, are largely dependent on the use of high technology and automation.

The new war-fighting doctrines

The new military operational concepts are to a large degree dependent on the successful exploitation of ET in the fields of advanced sensors, munitions, and delivery vehicles which have to be integrated within a complex command, control, communications and intelligence (C^3I) structure calling for automation and computerization for effective battlefield management. In the various new military doctrines the battlefield has been profoundly restructured. Although exhibiting differences in emphasis, they are more complementary than conflicting in that they stress primarily offensive-oriented deep strikes into the territory of the Warsaw Treaty Organization (WTO).

Deep strike

The emphasis on deep strike is reflected in the FOFA concept of the Allied Command Europe (ACE). It aims at a targeting capability for theatre-wide interdiction against key WTO forces and infrastructure with primarily conventionally armed cruise and ballistic missiles or airborne munitions up to more than 400 km beyond the 'forward edge of the battle area' (FEBA). This would cover the region from the German border to the Western military districts of the USSR. Under the deep-strike scenario of ACE the WTO would send its forces to the front in waves or echelons, with one echelon replacing the previous one as a complete unit, in order to cripple NATO's forward defences. By striking critical fixed targets (C^3 systems and choke-points such as bridges and transportation centres), by interdicting the forward movement of WTO reinforcing echelons while they are still deep in Eastern territory, and by attacking the armoured formations as they concentrate, ACE hopes to deny the WTO the ability to mass its forces successfully in the immediate battlefield area. With such a deep-strike option ACE expects to bring NATO's forward defence into manageable bounds. FOFA represents the 'classical' interdiction mission. The new element in this concept is that it seeks advanced conventional weapon capabilities to carry out interdiction missions previously inadequately covered or covered only by nuclear weapons. However, the new ET-based weapon systems, if available in sufficient quantities, could give the interdiction mission a new military quality.

The new US Army ALB doctrine as outlined in the 1982 version of the field manual FM 100-5, does not envisage attacks at the same depth as the FOFA concept because of its orientation towards the tactical and operational level of warfare. Throughout its foreseen area of operations ALB stresses generally offensive actions. This concept envisages an extended battlefield in which the corps level commander would attack uncommitted forces and support facilities as far as 150 km into WTO territory in order to create conditions to support offensive counter-attacks and defensive manoeuvre by ground forces. This war-fighting approach marks a fundamental shift in US Army doctrine. It replaces the previously more static- and attrition-oriented 'active defense' concept with one of mainly offensive-oriented manoeuvre warfare. ALB can be regarded fundamentally as a manoeuvre warfare doctrine which primarily seeks operational depth on the enemy's side of the battle area. Additionally, the ALB doctrine forsees the integrated use of nuclear, chemical and conventional weapons on its extended battlefield. A successful application of the ALB concept with its extended and integrated battlefield would require multi-corps and multinational co-ordination within NATO.[3]

Counter-air

Parallel to deep strike, a new operational concept for attacking airfields on WTO territory is also under discussion within NATO. A classified US document called 'Counter Air 90' proposes the employment of long-range ballistic missiles armed with runway-cratering sub-munitions to attack WTO main air bases. The objective is to keep the Eastern tactical aircraft from taking off or to deny their return to the same base, and to force them to use more vulnerable air bases for recovery. By destroying enough main WTO air bases it is hoped to reduce Eastern aircraft sortie rates considerably and to secure air superiority for NATO. According to its proponents, ballistic missiles are deemed necessary for these offensive counter-air missions because they alone are considered quick enough to ensure destruction of such time critical targets as air bases.[4]

Characteristics of the future tactical battlefield

A doctrine such as ALB is clearly heading for an automated battlefield in which highly centralized military staffs supported by various automated combat power procedures would directly wage the integrated air–land battle from something like an 'electronic hilltop'. The future NATO battlefield is regarded as one that: 'will be increasingly complex and so will the equipment. Unprecedented speed of action, lethality, variety of the threat, and sheer numbers of opposing weapons, systems and men make such complexity inevitable'.[5]

A future war in Europe is viewed by the military as likely to be a short and dynamic one. There would be vast amounts of information from various sophisticated airborne and ground-based sensors collecting threat intelligence data and reporting it to the military commanders as a basis for decision-making. Another important feature of future warfare would be the pressure upon the commander to decide and act rapidly because of new tactical weapons and highly mobile, offensive battlefield operations.

In view of the anticipated continued expansion in information flows, the growing need for quick decisions, and the need for flexibility, future warfare on the tactical battlefield is regarded by the military as posing threats and requiring actions too swift for human reaction and control. As a consequence, the realization of new operational concepts such as the ALB doctrine or FOFA depend largely on automation through computers. These computers have to meet operational demands such as real-time surveillance, accelerated information processing, real-time acquisition and priority selection of targets, rapid dissemination of mission orders, and quick weapon allocation and employment tasks. Computers will take over a large proportion of these tasks, which were previously carried out manually by soldiers. This could, for instance, be achieved through automated data-processing technologies which help collect, sort and categorize real-time battlefield information.

Current automation efforts on the tactical battlefield

The US Army, for example, has embarked upon a massive programme of battlefield automation in an attempt to realize General Westmoreland's vision. Its underlying, unspoken main aim is to ensure that the fragility of humans and their perceptual and mental limitations will be of less consequence in future war-fighting than they have been in the past.

Much of the effort of the US Army in battlefield automation is directed towards the tactical C^3I systems because these are the most critical, and hence most important, elements in future warfare. C^3I means integrating all the data from the different sensors and communication channels, determining their importance, and deciding upon and communicating the desired response to the units on the battlefield for military action. Success of the ALB doctrine is dependent on quick, reliable, flexible and, last but not least, survivable C^3I for the effective handling of the various elements of combat power on the different parts of the battlefield. The automation of tactical C^3I systems implies that decision-making by the commander will depend primarily on the use of computers. Systems being developed for battlefield automation of tactical C^3I in the US Army to wage the integrated air–land battle include the following:

1. The Joint Tactical Fusion Program (JTFP) which combines an Army system, the All Source Analysis System (ASAS), and an Air Force system, the Enemy Situation Correlation Element (ENSCE), to develop an automated tactical intelligence fusion and correlation system. Its goal is to help the tactical air and ground commander to correlate, integrate and interpret raw intelligence data with the minimum delay. Fielding of a small number of basic systems is planned to begin by 1987.[6]

2. Related to ASAS is the Maneuver Control System (MCS). It is an automated command and control system consisting of a number of microcomputers linked together in a local network, so that the available information can be accessed on any terminal. Local networking is of particular importance because the US Army intends to enhance survivability by dispersing its command and control functions. MCS gives the commander nearly instantaneous battlefield information, and once he makes a decision, the system facilitates the preparation and distribution of mission orders via millisecond bursts of digital transmissions to subordinate troop units. An initial development system for MCS has already been tested in Western Europe and initial operating capability is planned for 1986.[7]

3. Another element of the US Army's move towards tactical battlefield automation is the Position Location Reporting System (PLRS)/Joint Tactical Information Distribution System (JTIDS) hybrid. PLRS is a joint army–marine corps development. It is an automated digital communications network which intends to provide combat commanders with the capability for near real-time identification and location of their own and other friendly units on the highly mobile and extended battlefield. JTIDS, developed by the Army and

Air Force, automates tactical data distribution via a 'smart' message-routing capability, providing reliable, jam-resistant and survivable communications. PLRS and JTIDS will be merged to form the Army Data Distribution System (ADDS). Initial fielding of PLRS is scheduled for 1986, and a production decision on JTIDS is expected in early 1987.[8]

4. Tactical battlefield automation is also being extended to areas such as fire control, air defence, electronic warfare and combat service support.[9] NATO, for example, is developing on a multi-national basis a large automated command and control system for the deployment of its tactical air forces in Europe, the Airborne Command and Control System (ACCS). ACCS is intended first to complement the NATO Air-Defense Ground Environment (NADGE) and later to replace it. Its implementation is planned for the 1990s.[10]

5. Automation on the tactical battlefield is also being applied to weapon systems. A major effort is being directed towards the guidance and control of tactical missiles. This builds on and extends the breakthrough which was made with respect to the cruise missile in the 1970s. But even the 'classical' tank is seeing an increasing measure of automation. Until recently, automation extended only to the automatic stabilization of the main gun. The introduction of automatic ammunition loading systems for tank guns, which has already been realized in the Swedish S-main battle tank (a modernized version of the Strv 103B, to be known as REMO 103B), is a distinct step forward in the automation of this weapon system. Such automatic ammunition loading systems enable higher firing rates and an enlargement of the calibre of the tank gun beyond 120 mm. Owing to rapid progress in the field of digital signal processing, automated target acquisition and tracking systems are now envisaged as the next logical step in tank automation. The Tank Automotive Command of the US Army is already investigating such a possibility in its Multispectral Target Acquisition System (MTAS) programme on the basis of optronic and millimetre-wave sensors. Automated sensor-supported warning and countermeasure systems are another field currently under investigation for future tank technology.[11]

Problems of the automated battlefield

Modern computer technology based upon rapid improvements in data-processing hardware and upon, albeit somewhat slower, progress in software has enabled the development of advanced, complex battlefield automated systems. However, these technical breakthroughs also present severe deficiencies and problems for the military. Problems occur where the human exchanges information with the computer. The obvious problem is the adjustment of the human intellect and its mental processes (human short- and long-term memory, audio and visual sensory capacity, and information-processing speed) which risk being swamped by information and hence losing the overview when working with large, complex automated systems. Current concepts of battlefield automated systems, especially those for C^3I, obviously do not adequately address the limitations of human mental capabilities, as articles in military journals on the necessity of 'human factors integration' for these systems indicate.[12]

Extensive tactical battlefield automation, as is the case in the C^3I realm, is increasing both the speed and the volume of information inflows available to the commander. Thus, it is highly likely that too much information will be at his disposal. Having more (real-time) battlefield data than can be effectively analysed, decided and acted upon might adversely affect the commander's capability for instantaneous decision-making: an important requirement on the modern tactical battlefield. Doctrines such as ALB or FOFA

> are particularly dependent on quick, effective and accurate decisions based upon numerous operational factors from the disposition of enemy defences to natural influences such as terrain and weather, to the availability and disposition of friendly firepower and equipment.[13]

Artificial intelligence and the military

In order to solve the problems created by present and future battlefield automated systems the military is looking towards new computer processing technologies, and particularly a new software application tool called artificial intelligence (AI), in order to make the decision-making process more manageable for the commander on the future tactical battlefield. In conventional battlefield automated systems the computer is in no way a decision-maker. It can only receive, store and process data with incredible speed, that is, it merely prepares the decision-making process. AI aims to directly influence that decision-making process in a significant way by 'intelligently' automating it. In view of this possibility, the military faces the problem of deciding which elements of the commander's decision process on the tactical battlefield could be substituted and automated by some form of machine intelligence.

For the military AI also seems to hold the potential for intelligent automation of weapon systems which might be a further pressing requirement on the modern tactical battlefield. That is,

> time and situation will probably preclude use of systems that have an inherently lengthy decision process built into their application. Given the threat environment in the target area, for example, fighter aircrews cannot tolerate 'smart' systems that necessitate long exposure to hostile fire in order for the acquisition, lock-on, launch decision process to run its course.[14]

AI is the new exciting tool of computer science today. Despite widespread disagreement as to the definition of human intelligence—and what therefore constitutes machine intelligence—of all the future technologies it is AI that has become a new high-tech keyword for the military in the 1980s. Their hope is that a broad range of battlefield problems will be solved by means of AI. Besides coping with the increasing complexity of battlefield decision-making and military hardware, another main driving force behind the military's interest in AI is the issue of manpower. They hope to limit the adverse consequences of the projected declining availability of skilled manpower and of rising personnel costs.

To a large extent, AI has been boosted by the Strategic Computing Program established in 1983 (under the auspices of the US Defense Advanced Research Projects Agency—DARPA), the aim of which is to develop high-performance machine intelligence primarily for military purposes. According to the former director of DARPA the specific goals of the programme are as follows:

> While computers are already widely employed in defense, current computers have inflexible logic, and are limited to their ability to adapt to unanticipated enemy actions in the field. This problem is heightened by the increasing pace and complexity of modern warfare. The Strategic Computing program will confront this challenge by producing adaptive, intelligent computers specifically aimed at critical military applications. These new machines will be designed to solve complex problems in reasoning. Special symbolic processors will employ expert human knowledge contained in radical new memory systems to aid humans in controlling the operation of complex military systems. The new generation computers will understand connected human speech conveyed to them in natural English sentences as well as be able to see and understand visible images obtained from TV and other sensors.[15]

In the eyes of the military the potential application areas for AI on the tactical battlefield appear extensive. They range from surveillance, target acquisition, autonomous combat vehicles, navigation, multi-sensor fusion, terrain analysis, signal processing, weapons maintenance aiding devices, training devices, logistic support, image interpretation, tactical decision aids, war simulation to intelligent robotics. The field of applied AI which seems to have the prospect of a near-term military application involves what are known as expert or knowledge-based computer systems. An expert system imitates the judgement of human experts in a given domain of expertise. It is equipped with a knowledge-base of a human expert and approaches a problem in a similar way to the human thought process by machine reasoning and inferencing.

Possible applications of AI on the tactical battlefield

In this section a number of examples of potential military application areas of expert systems for the automated tactical battlefield are summarily outlined.[16] Some problems related to software and hardware as well as to the acceptance of this technology by soldiers are also addressed.

Tactical C^3I (TC^3I)

Intelligent automation of decision-making is regarded as of particular relevance to the TC^3I realm. It is intended to cope with the steadily increasing requirements for quick force interactions driven by new offensive-orientated operational concepts and by the expanding demands imposed by real-time battlefield surveillance information and sophisticated weapon systems of the friendly and opposing armed forces. Tactical commanders would have to make crucial decisions in the future under increasing time pressure in order to carry out offensive air and ground operations involving battlefield situations of great complexity. In the opinion of E. Taylor, director of requirements analysis of the TRW Defense System Group investigating AI applications for the US Army's offensive ALB concept, the future military TC^3I system (see figure 6.1) will 'depend on cooperative interaction between men and machines, and the interactions must be fast and comprehensive if the commanders are going to defeat a physically superior enemy by outwitting him'.[17]

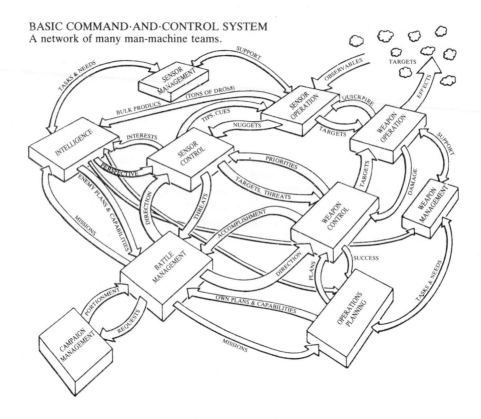

Figure 6.1. The future TC^3I system
Source: Redrawn from Taylor, E. C., 'Artificial intelligence in the Air–Land Battle', *Astronautics and Aeronautics*, July/August 1983, p. 57. *Courtesy of Aerospace America.*

Tactical command decisions aids

The use of expert systems as tactical command decision aids is one of the potential applications of AI which the military believes might overcome problems of conventionally automated TC^3I systems. This would be a so-called 'top-down' application in which expert systems were used by high-level command staff to manage the integrated air–land battle. Battlefield data analysis, situation assessment, and target and weapon allocation planning are among the functions that such a system might perform. Military expert systems of this type are already under development. Such a prototype expert system is the Tactical Air Target Recommender (TATR) for DARPA being developed by the RAND Corporation. TATR applies a knowledge-engineering problem-solving approach in order to help with planning and decisions concerning offensive counter-air missions in a wartime environment. It is based on decision-making techniques developed by experienced US Air Force tactical air targeters. Owing to this codified knowledge of military experts TATR is said to be capable of converting mission objectives into target recommendations and providing an attack plan. According to its engineers, TATR demonstrates a potential ability to assist the tactical air commander in the selection of and assignment of priority to enemy airfields, in determination of the targets to attack on those airfields, and in identification of the weapon systems that can most effectively be employed against those targets.[18]

Tactical command decision-aids to assist in conventionally automated TC^3I functions could be introduced to the armed forces in the near future. The development of this technology is regarded as sufficiently advanced to permit the intelligent automation of at least some functions. The main problem in the military deployment of these systems is not merely the technological one of engineering the laboratory prototypes for actual field use. Rather a chief obstacle seems to be the potential psychological problem of resistance to automated decision-making from military commanders. Parts of the military, especially the lower- and middle-level commanders, are already expressing their reservations about such systems. They fear that automated command decision tools might reduce them to something like electronic servitude. In their opinion, it is the human commanders that must still command, and this task can never be surrendered to a machine.[19] The AI developers in the military R&D community seem to be aware of this troop acceptance problem. A software engineer who worked for some time at the RAND Corporation on the development of AI-based military modelling tools spelt out the problem in the following way:

Machines that appear to threaten the autonomy and integrity of command cannot expect easy acceptance. It would be disastrous to introduce them by fiat. We should be studying how to design military management systems that reinforce, rather than undermine, the status and functionality of their middle-level users . . . we urgently need to address this question with the right mix of technological, political, doctrinal-legal, and command skills.[20]

Automated battlefield sensor fusion

'Bottom-up' applications in TC^3I systems are those which interface with the 'real' battlefield world in various ways: managing intelligence data, providing inputs to the higher command hierarchies and distributing messages. Automatic battlefield sensor fusion would be a key application in this realm involving a distinctly more complex process which would require very advanced expert systems. Sensor fusion is the correlation, merging and interpretation of the inputs from the various types of sensor distributed on the tactical battlefield (infra-red sensors, motion sensors, radar, etc.). This task demands knowledge on the observed military environment and the priorities for channelling the right information to the commander. The vast amount of real-time battlefield data available may very quickly overload the human military intelligence officer, even with today's conventional computer assistance. For example, 'a Warsaw Pact force facing a U.S. corp would produce thousands of sensor reports per hour. Intelligence analysts would have to observe and report significant changes within an hour to permit a commander to respond effectively.'[21]

Thus, the key limiting factor in real-time battlefield intelligence estimates would be the mental limits of the military analyst. Experiments conducted in the TRW Defense System Group with military intelligence analysts working in a one-hour time frame have shown that the quality of their assessments varied widely and that they often missed crucial information or gaps in the data. 'Even with a fast-response data-base-management system,' concludes one study, 'it is hard for any analyst under pressure to abstract good situation assessments from such large data streams.'[22]

Intelligent battlefield sensor-fusion systems based on advanced expert systems could process the raw intelligence data, employing stored knowledge on the optimal utilization of incoming information, and provide the human battlefield analyst with refined conclusions. The application of automatic battlefield sensor fusion could range from several sensors, connected only to weapon systems, to a whole theatre network. The TRW Defense System Group, as a prime contractor in a joint US Army, US Air Force and DARPA project, is developing a prototype knowledge-based tactical intelligence support system called the Battlefield Exploitation and Target Acquisition system (BETA). It is a precursor to a future ALB 2000 system to be developed in one of the US Army's high-priority R&D projects over the next decade—the Very Intelligent Surveillance and Target Acquisition (VISTA) project.[23] The aim of the BETA project is to automate sensor correlation, target generation, and situation display at Tactical Air, Corps and Division command level. Such a system would operate on a knowledge-based computer system to rapidly draw out, assign priorities and evaluate data from many different sensors. It would display easy reports of battlefield situations and targets for the intelligence analyst which are easy to survey. Furthermore, it helps to formulate military responses in the form of (near) real-time target nominations that aid, for instance, in battlefield interdiction (deep strike).[24] A similar support system for AirLand Battle intelligence management, called the 'Order of Battle Version 1 Knowledge Based (OB1KB)' system, is being developed by the MITRE corporation for the US Army to gauge the utility of AI in a tactical environment.[25]

In this context it is interesting to note that the military R&D work on automated battlefield sensor fusion for situation assessment seems to be benefiting from research work on civilian-orientated expert systems. According to the director of the BETA project a knowledge-based system for medical diagnosis under development at the University of Pittsburgh solves a problem very much like a military situation assessment because 'it examines incomplete data, constructs promising interpretations, tasks sensors, and iterates until it has a coherent solution'.[26] Yet the development of military expert systems for real-time battlefield sensor fusion to assist the military personnel in situation-assessment poses difficulties of great magnitude. Such a task requires advanced knowledge-based systems,

that first pick the right regions from vast and diverse knowledge spaces and then help operators converge to satisfactory solutions with just a few iterations. The challenge is much more than the difficult task of acquiring and representing the required knowledge. The challenge is developing a programmable description of the quick and ready insight that enables the very best operators to grasp the right data and right knowledge at the right time.[27]

Knowledge-based technologies are not expected to make an immediate contribution to any large-scale military application of this nature. In addition to the latter problem, automated battlefield sensor-fusion systems are also faced with the severe problem of particular vulnerability to enemy countermeasures such as jamming or spoofing. This fact is of special importance because the modern tactical battlefield will be characterized by a growing proliferation of electronic warfare systems and countermeasures for deception. For example, if NATO is to rely increasingly on automated systems for battlefield sensor fusion in order to carry out its FOFA concept, then the armed forces of the Soviet Union could adapt their war-fighting behaviour to the creation of a range of inputs which would be designed to match the decision-making heuristics in such an automated intelligence-gathering system. The Soviet Union could cause large perceptual mistakes in automated systems of this type by using electronic deception measures. Such a system acts as an information amplifier and would therefore amplify the disinformation created by the adversary. Canby, a noted critic of NATO over-reliance on emerging technologies, gives some examples of how automated battlefield sensor-fusion systems might be badly fooled:

False images can . . . be created to deceive the sensors and swamp and break down the VISTA system. Formerly false images on the battlefield were expensive. . . . Now they can be . . . created by deceiving 'simple-minded' sensors and simulated electronically across a number of sensor modes. For example, the MTI radar detection system can be deceived and the entire system overloaded by the simple expedient of sending civilians or soldiers with corner reflectors on their caps jogging or bicycling in single file from town to town and forest to forest, thus appearing as innumerable columns of vehicles moving from hide to hide. In one known Soviet simulation technique, vehicles are made to appear to be moving down roads, when in fact there is nothing more than a line of (small and difficult-to-hit) radar reflectors (to overload signal processing) and emitters (to attract fire) strung parallel to a road like Christmas-tree lights.[28]

Although Canby's argument is relevant, one wonders if soldiers would have the time and opportunity to jog and cycle on the future tactical battlefield.

Distributed battlefield knowledge-based system

Another important element of a TC^{3}I system is the communications network which distributes mission orders and battlefield information. This tactical communications network needs management in order to distribute information and to reconstitute itself in a flexible way in case of damage. Automation might be applied here for monitoring and reconfiguring the availability of military communications channels during the course of war-fighting. In its R&D programme for the battlefield of the 1990s the US Army has set up the Distributed C^{3}I (DC^{3}I) project which aims at fast, flexible, jam- and intercept-resistant continuous battlefield-information transmissions under adverse war-fighting conditions.[29]

Nevertheless, a battlefield communications network provides more opportunities for the military than a mere 'advanced telephone system'. It may be used to constitute a distributed battlefield knowledge-based (DBKB) system and so create a capability for distributed tactical battlefield problem solving: 'By connecting various computerized stations it becomes possible to solve problems using knowledge and information from a group of sites'.[30] A first step towards such a DBKB system might take the form of interfacing the various tactical command decision-aids so that they may exchange battlefield information and problems. Such a modular approach would leave the configuration of the local systems with some measure of independence. A DBKB system of this type is regarded as feasible in the near future. A much more demanding task would be the design of a single fully DBKB system using all the multiple computers, battlefield sensors and knowledge bases. With such an advanced system the 'processing burden could be distributed dynamically in response to shifting loads, facility losses, and communication link outages'.[31] However, the demands on the required knowledge-based technologies for this application are enormous, involving great development risks.

Intelligent tactical warfare simulation systems

In contrast to numerical simulation, knowledge-based programming opens up new possibilities and application areas of military simulation which have hitherto been too complex to grasp. Simulations on the basis of expert systems allow one 'to model the logical connectedness which is common in the structure of human institutions, artefacts, and behaviour'.[32] Intelligent military simulation systems embody a military expert's knowledge of the battle objects comprising the given military simulation domain. Such systems could be used by the commander as an intelligent decision-making aid on the tactical battlefield enabling him quickly to analyse and evaluate various possible outcomes of, for example, different offensive manoeuvre warfare options. A prototype expert system of this type, called TWIRL, for tactical ground-based combat simulation, has already been developed by the RAND Corporation.

The TWIRL system simulates a ground combat war-fighting situation on the tactical battlefield between two opposing armed forces including troop deployment, artillery firing, air interdiction, and electronic communication and jamming. The TWIRL simulation model could also be used to investigate electronic warfare issues.[33]

Weapon system fault diagnosis and maintenance expert systems

The increasing complexity of today's weapon systems aggravates the problem of their reliability and repair. Automating the fault diagnosis and maintenance process of weapon systems through knowledge-based systems is regarded by the military as another useful application area due to the limited number of skilled technicians and maintenance personnel available. The time required to repair and maintain crucial and expensive weapon systems is seen as a very important factor for the fighting capability of the armed forces on the tactical battlefield. In particular, the repair and maintenance of electronic and mechanical components of tactical aircraft is considered an important application area for such expert systems. The automated process might take the following form:

the aircraft mechanic will approach the broken aircraft with a terminal in hand rather than reams of tech manuals. The technician will input what is known about the aircraft's failure, and the computer will initiate an interactive question-and-answer series with the technician—the technician supplying answers to the computer's questions. The computer will identify the problem and tell the technician how to fix it.[34]

Research on hardware fault diagnosis is already being conducted, but fault diagnosis is a highly skilled task which is difficult to codify in a knowledge base. The possibility of realizing very intelligent systems for this task in the near future seems unlikely. What might emerge soon are expert systems with limited functions which ease the tasks of weapon-system maintenance personnel. The automation of some steps in this task is regarded as immediately feasible. An expert system of a primitive type for weapon system maintenance could, for example, indicate which pages of the weapons' electronic system manuals are likely to be related to the external symptoms.[35]

Automated assistance in the use of tactical weapon systems

The military expects to reduce the workload of the human operators of increasingly complex weapon systems by incorporating computers on board those weapon systems which run knowledge-based programs. Such expert systems are designed to take over those tasks for which the human weapon operator has too little time and thereby help him to concentrate on the most important tasks in tactical mission execution. One possible application of this kind is the intelligent pilot assistant developed by DARPA in its Strategic Computing Program. The expert system would possess knowledge on the operational modes of the fighter plane and hence support the pilot in tactical mission planning and execution. Such a system would assist the pilot and his decision-making process in a number of ways, for instance, aid in tactical mission planning in real-time en route, in determination of employment tactics for mission execution, in the takeover of responsibility for the operation of the many sub-systems on board, in the surveillance of internal and external sensors on the aircraft, in giving aircraft status reports and continuous mission situation assessments, and in informing the pilot in natural language on the latter and responding to voice control.[36]

Another possible application of this type for the tactical battlefield is being investigated in the Ninja research programme at the RAND Corporation. The goal of this research project is the development of an advanced, highly automated, survivable, light ground-combat vehicle system: the future 'high-tech' tank. The Ninja concept, which builds mainly on AI and robotics technologies, aims to raise markedly the combat capability and effectiveness of US Army light division forces and US-NATO armour forces on the tactical battlefield. AI technology holds out the prospect of significantly reducing the size of the tank without proportionally diminishing its combat effectiveness. This might be possible by shifting many complex, interrelated tasks—which currently require several human tank crew members—to automated sub-systems via expert systems. Thus, a single human tank crew member might take over the remaining, largely supervisory and decision-making orientated tasks (see figure 6.2.). Major features of the Ninja concept include integrated multiple sensors for fire control and target acquisition, integrated multiple

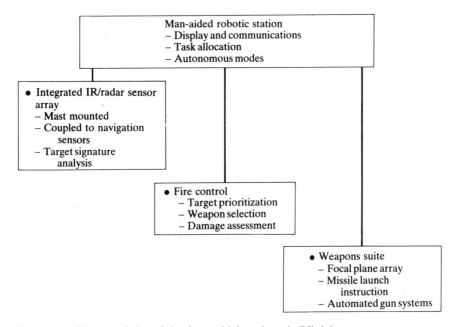

Figure 6.2. Characteristics of the future high-tech tank (Ninja)
Source: Clark, J. R. *et al.*, *An Introduction to NINJA*, Rand note N-2222-ARPA/RC, Oct. 1984, p. 11.

weapons (missile, gun, machine-gun and grenade launcher, etc.), tactical and strategic transportability, high survivability due to small vehicle size, high mobility and augmented armour protection.[37] Such possible applications of advanced expert systems are seen as potentially attainable within the next decade.

Other AI applications on the tactical battlefield

A number of AI disciplines which do not fall under the category of knowledge engineering are of great interest to the military. These include machine vision and robotics. In these fields fundamental non-knowledge-based issues must first be solved. Progress in two generic technologies is required as the basis for machine vision and robotics: electronic devices and image processing. Military R&D in electronic devices aims at the development of advanced sensors for navigation, surveillance and target tracking. This includes, for example, a near real-time 3-D imaging sensor. Military research in image processing (for which the military has a real-time requirement) is concerned with extracting target data from sensor output and sorting out a target's outline in the face of background clutter and enemy countermeasures.[38] Once such progress has been made, then potentially important application possibilities of knowledge-based systems arise: an image-interpretation expert system which interprets the information from the advanced imaging sensors and produces a symbolic description of the battlefield environment and objects within.[39]

Machine vision/image interpretation   Image interpretation by computer—for example, the analysis of digital information aimed at understanding the content of an image—holds great military potential. An AI-based vision system is required not only to recognize different objects, but also to understand the relationship of the objects to one another. Potential military application areas of machine (computer) image-interpretation technology on the tactical battlefield also include robotics (see below), battlefield aerial-survey image interpretation, target identification and classification, and weapons guidance.

For the military R&D community, one very interesting possible application for this branch of AI technology is in missile guidance, that is, to create a highly automated, intelligent tactical ('fire and forget') missile containing something comparable to an 'autonomous nervous system and brain'. Such autonomous tactical missile homing devices, when equipped with high-speed, real-time image-processing technology knowledgeable about tactical target parameters, could reach the target area without any external support (such as radar or laser designation), and then autonomously detect, identify, discriminate among and attack discrete high-value targets (tanks) on the battlefield (see figure 6.3).[40] According to one report, the Hughes Missile System Group has been investigating AI-based image-processing technology on its WASP anti-armour missile project which prior to its cancellation was a part of the Wide Area Anti-armor Munitions (WAAM) programme of the US Air Force.[41]

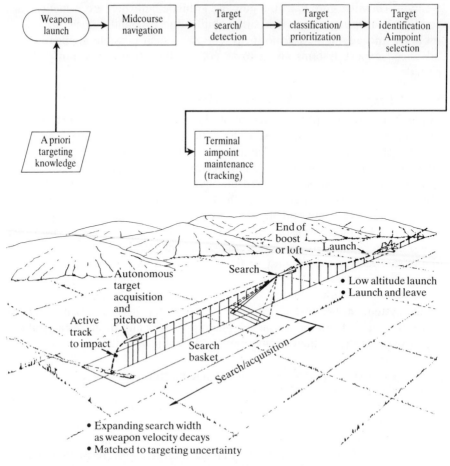

Figure 6.3. The autonomous, 'intelligent' tactical missile
(a) The basic algorithm sequence for a fully autonomous, 'intelligent' missile
(b) The autonomous stand-off LOAL acquisition and tracking procedures.
Source: Redrawn from Richardson, K., 'Tactical missile technology for the 1990s', *Military Technology*, vol. 10, no. 10 (1986), pp. 218 and 221.
Courtesy of Hughes Aircraft Company.

Development of machine vision will have to overcome many difficulties. Current image-interpretation systems are only able to identify objects in a simple, uncomplicated structured environment. Very advanced machine image-interpretation necessary for complex scenes such as the tactical battlefield will require a large measure of 'common-sense' knowledge about the real physical world. In image processing substantial problems have to be solved in the extraction of features from the raw image.[42] Taking into account the real-time requirements on the future tactical battlefield, some scientists in the military R&D realm are sceptical as to whether machine-image vision analysis can achieve an accurate target description in real-time, 'such as distinguishing between a T-72 tank and a M1 tank at a distance'.[43]

Robotics The military sees great prospects for robotics in tactical battlefield applications. If this technology could be sufficiently developed, it would allow the creation of autonomous and semi-autonomous weapon systems. Progress in the military uses of robotics in the near future will depend on developments in advanced sensors and image processing (machine vision) and knowledge-based technologies for situation assessment and planning.[44]

The US Army, whose new ALB doctrine puts primary emphasis on mobility, firepower and electronics, is highly interested in AI-based robotics tanks or fighting vehicles that could perform autonomous unmanned attack, reconnaissance or supply missions on the battlefield. Such intelligent robotics tanks operating autonomously on the battlefield could be used, for example, as the first attack echelon in offensive deep-strike manoeuvre warfare operations. In its Strategic Computing Program, DARPA is investigating this possibility through the Autonomous Land Vehicle (ALV) project. This project aims at a technology demonstration vehicle that can operate independently and intelligently in complex battlefield terrain. Having received general mission instructions, the vehicle would execute a navigation plan by using its vision system and image-interpretation and planning expert systems to accomplish its mission.[45] Intelligent combat vehicles are still far from being realized in the near future because the computational requirements of such AI-based systems exceed the capacity of today's most powerful computers by almost a thousandfold.[46]

An intermediate goal which might—according to the military R&D community—be feasible in the foreseeable future would be the development of semi-autonomous, remotely controlled robotics combat vehicles or weapon systems equipped with a certain measure of AI.[47] One such possible application might be, for instance, an unmanned, intelligently automated howitzer which would independently carry out ballistic computations and fire on the target after having received target information from a crew on board a control vehicle stationed safely behind the forward battle line.[48]

Problems in the application of AI technology to the military domain

The military is looking towards AI technologies as a so-called force multiplier to extend the power of existing unintelligent military computers and weapon systems. Furthermore, they expect to augment and eventually even to replace human military problem-solving with AI-based military systems assuming greater operational responsibility on the tactical battlefield. But is this technological vision of an intelligently automated battlefield a sound one? Important gaps between basic and applied AI still have to be bridged (for example, the nature of human versus machine intelligence). However, even applied AI in the form of knowledge-based systems is still a relatively untested and immature technology.[49] So far, there are no applications of expert systems in operational military systems. Effective military applications are a long way from being demonstrated, let alone proved. This is particularly the case for automated battlefield intelligence analysis systems, 'where the connectivity of millions of bits of seemingly unrelated information must be recognized by the system'.[50] Even AI experts working on military R&D applications caution: 'As with most new technologies, the expectations of their eventual impact can be overestimated. To some extent that has happened with the premature build-up of artificial intelligence.'[51]

So far, expert systems developed for the tactical battlefield and other military domains have been only laboratory models. Tactical command decision aids already exist in prototype form and have been demonstrated, but it has yet to be shown that they can be engineered for and implemented in an actual operational battlefield environment. Very substantial knowledge-engineering problems have to be solved in order to realize this technology for military applications. One major problem seems to be the construction of a trustworthy knowledge base (rules) for a trustworthy battlefield environment covering a realistic spectrum of military situations and details on the tactical battlefield. The design and development of comprehensive knowledge bases for actual battlefield use is crucial. Yet so far,

All of the AI systems that have been built have dealt with honest environments. No one has deliberately falsified the information provided to any of them. Even without deliberately false information, coping with the uncertainty of both the raw input information and propositions deduced from it represents a substantial challenge.[52]

The record in conventional warfare experience so far suggests that intangible factors such as uncertainty and surprise play an important role on the tactical battlefield.[53] It remains to be seen if it is at all possible to codify the important battlefield factor of uncertainty realistically in a military expert system. This problem is aggravated if one considers the probable widespread use of electronic deception measures central to future battlefield war-fighting. In view of these substantial problems it is reasonable to assume that complex knowledge-based systems—which are required especially by integrated C^3I systems on the battlefield—will make errors. They might even break down completely because expert systems 'can only effectively deal with a uniform knowledge base. If there are any conflicts in the knowledge base ... the system will collapse'.[54] These systems will in all probability show 'flashes of brilliance' in one situation and 'incredible stupidity'—e.g., false recommendations or decisions—in another. Such probable errors, very difficult for the military commander to recognize, entail military repercussions as well as potentially grave political costs given the extreme time pressures on decision-making on the modern tactical battlefield. This political problem is addressed in section III.

Other substantial problem areas in the development of expert systems include: the knowledge-acquisition process crucial for the construction of a knowledge base that could be a potential bottleneck in the future; critical shortage of skilled personnel (programmers, system designers), especially with military problem-solving experience; and software engineering costs.[55]

On the whole, 'successful' military application of AI on the future NATO battlefield with its (near) real-time requirements depends on the realization of fifth-generation computer architecture permitting parallel processing and very high execution speed 'for making time critical decisions that lack detailed knowledge' as well as associative memories in which data are stored and indexed by content.[56] The military's projected reliance on AI-based computer systems for the next generation of automated C^3I systems and autonomous quasi-intelligent weapons is very much based on future advances in microelectronic hardware. Such advances are above all expected from the US DoD VHSIC programme, a high-priority technology effort under military control.[57]

III. Command and control of tactical nuclear weapons

The military also sees opportunities for AI applications in the field of command and control for tactical nuclear weapons. By means of this technology it is hoped to introduce more flexible nuclear war-fighting options on the tactical battlefield.

There are thousands of tactical nuclear weapons intended for battlefield use in regional conflicts, such as a war in Europe between NATO and the WTO. Tactical nuclear weaponry includes short-range nuclear missiles, nuclear artillery shells, and other nuclear devices.[58] Both the major global powers, the USA and the USSR, have constructed a complex technological apparatus for command and control of their nuclear weaponry. However, the organization and management of command and control for tactical nuclear weapons, as opposed to strategic weapons, occurs in a different military force structure setting posing special problems. Referring to the structure of the US nuclear forces, Bracken states:

The strategic nuclear forces of the US are as safe and controllable as they are because the command organizations which govern them have been designed in and for the nuclear age. These organizations were built around the nuclear weapons they were designed to control. This is not the case for tactical nuclear weapons, which, by reason of necessity, have been built into a general-purpose force structure of corps, divisions and brigades. Command and control, especially of a political variety, is very different because of the radically different institutional ethos of the Army compared to SAC.[59]

The term 'different institutional ethos' suggests that US/NATO armies have not been built around tactical nuclear weapons but, on the contrary, tactical nuclear arms have been implanted in traditional World War II-style military organizations. Additionally, in NATO there is also the problem of integrated command and control of nuclear weapons. Between 1956 and 1961 the US Army undertook the unparalleled attempt to reorganize its force structure for operations on the tactical nuclear battlefield: the development of the so-called Pentomic division. With the advent of the Kennedy Administration and its political emphasis on a more conventional defence-orientated military posture for NATO the Pentomic division structure was abandoned.[60]

Subsequently, although the USA deployed thousands of tactical nuclear weapons, designed for war-fighting, on West European soil during the 1950s and 1960s, doctrinal questions of tactical nuclear warfare were scarcely a topic of discussion in NATO. The adoption of flexible response by NATO in 1967 provided a compromise formula which was ambiguous enough to allow all members to interpret the issue of tactical nuclear warfare as they preferred. The USA and the West European allies stressed the importance of tactical nuclear weapons, but for opposing interests: in the event of a European conflict, for the West European members these weapons represented the best guarantee for threatening the USSR with the risk of escalation to a general nuclear war, whereas the USA hoped by deploying such weapons to limit a war to Europe. These fundamentally differing political interests 'foreclosed any attempt to derive from declaratory doctrine precise guidance concerning the way in which theater nuclear weapons might ultimately be employed'.[61] Thus, until the end of the 1970s related organizational and management problems for command and control of nuclear war-fighting on the battlefield remained largely officially unaddressed in NATO. Furthermore, 'the requirements of the Vietnam war so preoccupied U.S. planners and doctrine makers during the late 1960s and early 1970s that a hiatus developed in planning for tactical nuclear warfare in Europe'.[62]

In their 1978 study on tactical nuclear weapons, Van Cleave and Cohen gave a harsh assessment of US and NATO armed forces in characterizing them as being 'ill prepared and fundamentally untrained for nuclear combat. Serious training of the U.S. forces for a nuclear conflict environment ranges from less than comical to non-existent'.[63]

The integrated battlefield

In 1979 the US Army took up the challenge and set out to review its war-fighting doctrine and tactics and to determine its future needs for fighting on a conventional, nuclear, and chemical battlefield, merging doctrinal ideas which had already been discussed. With the ALB doctrine the US Army upgraded its own nuclear and chemical war-fighting posture. A main element of the ALB doctrine is the central sub-concept of the 'integrated battlefield' in which deterring nuclear war gives way to the goal of fighting and winning a limited nuclear war. The rise of the integrated battlefield was based on the perception that the armed forces of the USSR are ready, willing and able to use nuclear and chemical weapons at the outset of any military conflict. There might be some indications for such a Soviet nuclear posture. Frei points out that for the USSR tactical nuclear weapons are: 'a deterrent, but, in case war is inevitable, a means to "clear an avenue for a massive offensive." ... By contrast to Soviet doctrinal statements, the Soviet capabilities so far deployed point to preparations for waging a limited nuclear war.'[64]

However, it would be politically misleading to deduce from this possible nuclear posture an image of a Soviet leadership intending to play something like a nuclear 'High Noon'. On the contrary, the USSR has, especially with regard to nuclear weapons, so far shown a crisis behaviour which is characterized by risk aversion and caution.

On the basis of its perception of the Soviet military posture the US Army developed an operational battlefield concept that, for the first time, integrated the tactical nuclear option and the deep second-echelon interdiction option into a general concept of tactics and operational manoeuvre warfare to secure military and political objectives in an aggressive way. The tactical use of nuclear weapons was regarded as providing an opportunity for initiating offensive operations on the battlefield.[65]

Requirements on the integrated battlefield

In the integrated battlefield the military planners were looking for more than just the option of using tactical nuclear weapons: their concept was one of integrated conventional and nuclear fire support with integrated manoeuvre warfare, and integrated air–land operations. The implementation of such a war-fighting concept would place enormous demands on command and control because planning, co-ordination and employment of nuclear weapons would have to be integrated on a continuous basis into ground manoeuvre warfare operations. It would mean that 'commanders had to take a whole-battle view, plan for preemptive enemy strikes, and develop an engagement strategy. . . . Command, control, and communications had to be organized to withstand a massive enemy strike.'[66]

Possible AI applications

From the military viewpoint such offensive limited nuclear war-fighting options for the battlefield—if they are to be carried out in the demanded flexible and timely way throughout the foreseen spectrum of conflict—require the automation of the nuclear force planning and employment process. This implies the development of an automated, comprehensive and survivable command and control capability where opportunities for military applied AI could arise. Militarized expert systems could offer more flexible nuclear war-fighting options for the tactical battlefield now demanded by the military, 'by using a predetermined set of rules and doctrines rather than a predetermined set of specific actions'.[67] The military hopes that the use of expert systems as a tactical nuclear command decision aid could significantly improve their ability to develop flexible tactical nuclear contingency plans, to modify those plans quickly or to develop entirely new nuclear battlefield plans in response to rapidly changing situations during a crisis or the course of war-fighting. Additionally, such a system could allow the commander to assess quickly the potential military consequences of executing various tactical nuclear options, and to plan and execute tactical nuclear strikes against new, suddenly emerging, enemy targets.

Automated nuclear weapon planning tools

According to statements made by the director of the US Defense Nuclear Agency (DNA) before the US Congress, an R&D programme aiming to provide US and NATO theatre nuclear warfare planners with automated nuclear weapon planning tools has been under way for some time. One development effort of the DNA in this regard is the Target Analysis and Planning (TAP) system. It is an 'automated aid, designed to rapidly assist in attacking second echelon targets with specific nuclear weapons in near real time' and 'is currently in use with V and VII Corps in Europe'. The US Army intends to incorporate TAP as a major part of a much larger planned integrated command and control system. The DNA has also initiated a similar research effort to provide automated nuclear planning aids for the US and NATO Air Forces.[68]

Another DNA development effort is an automated force-on-force simulation model which 'will enable analysts within army staffs to investigate current or new weapon systems and tactics as they would influence the nuclear and conventional battlefield'.[69] Such a simulation model could also allow commanders to rapidly and effectively analyse their conventional and nuclear plans and tactics in the light of the actual battlefield situation which they face.

Survivable command and control facilities

Nevertheless, the whole integrated battlefield concept is entirely dependent on a flexible and survivable C^3 capability. For the integrated battlefield loss of command and control poses a great risk of enemy nuclear and chemical strikes. C^3 within US/NATO corps, as configured today, is considered ill-suited to the demands of the integrated battlefield. Consequently, efforts to automate the planning and employment of tactical nuclear weapons are paralleled by intensified efforts to enhance the survivability of nuclear command and control centres. Measures envisaged by the military include providing redundancy by using multiple and highly mobile command centres which are also hardened to withstand nearby nuclear and chemical strikes. Flexible and effective communications systems on which the command centres depend for battlefield information and exercising control of tactical nuclear weapons are also needed. This could be achieved through a distributed communications system and through hardening the equipment to cope with the electromagnetic pulse (EMP), and gamma and neutron radiation.[70]

The question of political command and control

Automated nuclear command and control systems based on expert systems could promote the dangerous illusion that military commanders and political decision-makers could control a regional limited tactical nuclear war-fighting situation which, in fact, might very well quickly escalate into an all-out nuclear war. According to Ball, who has made a detailed study of this issue, controlling a nuclear war is not possible.[71] Furthermore, it is highly questionable whether the USSR makes the same distinction as does the USA between tactical and strategic nuclear war.

The integrated battlefield is based on the premise that nuclear weapons can be integrated with tactical 'campaigns', but, fortunately, no one has as yet any experience as to how military operations on the battlefield would actually proceed in the chaos of nuclear and chemical debris, or how the soldiers would react to such a situation. Furthermore, how can the delegation of initiative and offensive operations with tactical nuclear weapons be made compatible with the requirements of political command and control?

The military regards its perceived need for a credible and integrated tactical nuclear employment option as inconsistent with the present highly centralized *political* command and control system and the associated tactical nuclear release procedures of the National Command Authority. It believes that for the execution of its new battlefield concept the release authority for the use of tactical nuclear weapons would come too late: 'By the time a commander could clearly demonstrate the time to be right for use of nuclear weapons, it would already be too late. Commanders could not afford to plan and prepare for nuclear-dependent manoeuvre operations unless release was assured.'[72]

As portrayed graphically in the US Army *Field Manual FM 100-5* of 1976 the tactical nuclear release sequence could take as long as 24 hours.[73] In order to be 'successful' on the integrated battlefield the military demands that nuclear weapon release be 'responsive' to the battlefield, and that the release procedures be simplified. This would amount to a kind of pre-delegation of authority for use of tactical nuclear weapons from the president to the military commander in the field.[74] It has been reported that US Army officials informed a group of US Congress members in 1982 at a classified briefing on their ALB doctrine that they were seeking 'pre-clearance' for the use of tactical nuclear weapons in the Central European battlefield of the future so that field commanders could use their nuclear arms right at the beginning of a conflict, without having to wait for the time-consuming presidential authorization.[75]

IV. Escalation risks of new war-fighting doctrines for the automated battlefield

The technological and doctrinal trends inherent in the new Western military operational concepts raise questions concerning their escalation risks with regard to the nuclear threshold, their effects on crisis stability, and their implications for political decision-making in a crisis. These problems are addressed below.

The nuclear threshold: will it be higher?

Systems to support conventional deep-strike operations, currently under development or on the drawing board, are presented by the military and political leadership as an improvement in NATO's conventional defence capability and as an increase of the nuclear threshold in order to strengthen the credibility of the Western Alliance's strategy popularly known as flexible response (MC 14/3).

However, despite the declared intentions of improving the conventional capability, General Rogers has explicitly stressed that the Western Alliance generally cannot renounce the option of first-use of nuclear weapons—an escalation threat which could ultimately force the WTO to abandon a military attack and retreat.[76] When considering the new Western war-fighting concepts and the nuclear threshold, one should keep in mind that the ALB doctrine incorporates the integrated battlefield which could mean deliberate escalation to the nuclear level right at the beginning of a conflict if release were to be granted by the National Command Authority (see section II). This US doctrine regards nuclear weapons in the first place as military war-fighting instruments, not as a *political* means for war prevention or termination. Implementation of the integrated battlefield approach in NATO would mark a shift towards an offensive-orientated nuclear war-fighting posture with broad and undesirable ramifications for deterrence stability and intra-alliance relations. Consequently, this aspect of the ALB doctrine has led to heavy criticism in Western Europe, especially in FR Germany. NATO's Supreme Allied Commander, Europe (SACEUR) General Bernard Rogers was forced to declare officially that the integrated battlefield concept of the ALB doctrine is not compatible with the current strategy and doctrine of the Western Alliance.[77]

The notion that the proposed emerging technology-based non-nuclear weapons with their precision-guided sub-munitions would reduce escalation risks and render the employment of nuclear weapons less likely in a military conflict—thereby solving many of the dilemmas inherent in NATO's strategy and posture—is in the opinion of the author very questionable. The employment of conventional deep-strike systems 'may appear more attractive than present or proposed theater nuclear systems because they suggest the possibility of achieving similar effects in battle without the danger of escalation'.[78] On the contrary, the employment of conventional deep-strike systems as foreseen in the FOFA concept could very well result in undesired escalation. It could lead to a situation which Posen has called 'inadvertent escalation', as a path to nuclear war.

According to Posen this form of escalation would be the unintended consequence of a decision to conduct a conventional war in which escalation arises out of intense conventional conflict. Usually, escalation is considered to be a deliberate (rational) policy choice, in which the political leadership would decide to escalate a military conflict in the face of a possible conventional defeat, or an accident.[79] The possibility of inadvertent escalation will be influenced by choices with regard to operational plans and force postures. Posen mentions three factors as causes of inadvertent escalation: (*a*) offensive military operations; (*b*) ambiguities in offence and defence in connection with weapon technology and geography; and (*c*) C^3I problems arising out of the 'fog of war'.[80]

However, it seems to the author that the problem of escalation control has not particularly influenced the shape of the new Western war-fighting doctrines. The offensive operational preferences, for instance, of the FOFA concept for conducting a conventional war with deep-strike systems may 'unleash enormous, and possibly uncontrollable, escalation pressures despite the desires of American or Soviet policymakers'.[81] Yet, while the option of employing conventional weapons at the beginning of a war might delay NATO's recourse to tactical nuclear weapons, the use of conventionally armed long-range missiles could increase the likelihood of Soviet employment of nuclear forces for the reasons below.

First, the USSR would most probably be unable to distinguish between non-nuclear and nuclear missiles. If NATO's weapon plans are carried out, a large number of cruise missiles would be deployed on the battlefield, some equipped with nuclear warheads and others not. If the Pershing-II ballistic missile is engineered to carry a conventional weapon load, thus becoming a dual-capable system, the problem of discrimination for the USSR will be even more pronounced.

Furthermore, deep strikes by dual-capable missiles could put Soviet nuclear forces on the battlefield at risk because some systems would probably be targeted, for example, on command and control centres responsible for Soviet tactical nuclear forces or on the delivery systems themselves. NATO preparations for launching non-nuclear deep-strike missiles could be regarded by the Soviet Union, because of the distinction problem, as an escalation step to the nuclear level, thus forcing the Soviet military and political leadership into the early alternative of considering nuclear pre-emption or launch-on-warning before their own nuclear systems were destroyed. Essentially, such a situation would deteriorate with the increasing range (greater target coverage of important military assets), shorter flight-times (ballistic missiles), and increased accuracy of dual-capable missiles.

The 'use them or lose them' dilemma, a formulation commonly used to characterize the dangers of NATO's nuclear posture, would then also extend to the WTO. Thus, by blurring the distinction between nuclear and conventional weapons, the new deep-strike arms technologies may actually lower the nuclear threshold. An early large-scale use of nuclear weapons would become more likely and could lead to uncontrollable escalation. With their emphasis on offensive deep strikes by dual-capable missiles, the new Western war-fighting doctrines threaten to markedly accelerate the escalation process, a result which runs counter to the original purpose.[82]

Furthermore, if conventional deep-strike systems could carry out missions and reach a lethality equivalent to theatre nuclear weapons, they could also endanger vital Soviet strategic force installations (early-warning radars, air defence systems, strategic command and control facilities, and so forth).[83] Because of the geographical difference, the USSR does not dispose of such an option with theatre systems against the territory of the USA. It is difficult to believe that any Western attack on such strategic facilities with conventional weapons equivalent in overall effectiveness to a nuclear strike would not trigger an extreme Soviet reaction, for example a nuclear response.[84]

Thus, the assumption—inherent in the new operational concepts stressing deep strikes—that the USSR would abstain from escalation if such attacks were to be carried out 'only' with non-nuclear means and that NATO would maintain escalation control appears to be a very dangerous myth. It seems, in fact, that the technological and doctrinal trends emerging in the USA and NATO are reinforcing each other with the net result of systematically eroding the nuclear 'firebreak'. In the presence of such war-fighting concepts and weapons a future war would probably face the increasing risk of escalation across the full conflict spectrum into an all-out nuclear war.

Crisis stability

If military means are to serve the political purpose of making a war unlikely, military doctrines and the operational and weapon requirements resulting from them must be measured against the important political criterion of crisis stability: 'Crisis stability is a situation in which neither side forgoes crucial options and waits instead of delivering a pre-emptive strike. Under conditions of crisis instability, on the other hand, there are strong incentives to strike first if the occurrence of war is believed as a real possibility.'[85] If the goal of crisis stability is to be advanced then the advantages each side could gain from military prevention or pre-emption measures must be reduced as much as possible.

However, FOFA and especially ALB are clearly war-fighting doctrines with an offensive orientation since the new US/NATO operational concepts 'focus not on simply creating an attrition threat to enemy forces, but on the capability to defeat his operations and objectives and support a more "offensive" NATO defense'.[86] If NATO attained a realistic offensive capability, then both the WTO, which for historical reasons has long adopted an offensive-orientated military doctrine, and the Western Alliance would consider offensive operations in a crisis. Thus, it is quite probable that both sides would seriously contemplate a first strike in a crisis because each would be aware of the other's offensive capabilities. This would lead to pressures for pre-emptive measures in order not to lose crucial military options. The result would be a highly unstable situation reinforced by the new weapon technologies. The projected dual-capable weapons for deep-strike operations—capable of being used in an offensive manner—would, in the context of their operational employment concepts, create instabilities which were formerly only associated with false developments at the strategic level.

The growing proliferation of deep-strike weapons on the battlefield implies that critical military facilities and weapons of both the WTO and NATO will be increasingly vulnerable. According to the ESECS study NATO would require at least 900 non-nuclear missiles for counter-air (attack of air bases) and interdiction (choke-points) missions. Furthermore, another 3000 ground-launched corps support missiles and 2000 air-delivered missiles with terminally guided submunitions for battlefield interdiction of follow-on-forces or destruction of air defence systems are regarded as necessary for a credible non-nuclear deep-strike option.[87] If NATO is developing extensive deep-strike capabilities through the use of precision-guided non-nuclear missiles with ranges of about 400–500 km, then the WTO will probably be compelled to copy such a development complementing their existing theatre nuclear missile inventory with conventional ones.

The availability of non-nuclear deep-strike weapons could increase the likelihood and the feasibility of pre-emption (because such weapons do not seem to carry with them the politico-psychological barriers which might inhibit the employment of nuclear weapons):

What one sees developing is a situation in which all crucial fixed assets of each side within several hundred miles of the border are at risk, and could be destroyed within minutes of an outbreak of hostilities. . . . In such a situation, the disastrous consequences of waiting too long before launching deep strikes must be on the mind of every commander.[88]

The pressure for timely military action during a crisis is a clear indication of loss of crisis stability. If the pressure to strike first is so overwhelming it could mean that military considerations prevail over the primacy of politics. The decision to strike deep at the adversary's follow-on-forces or airfields will in all likelihood have to be taken at an early and probably still ambiguous stage of the conflict. A case in point is the offensive counter-air mission connected with the ballistic missile option because it involves particularly destabilizing tendencies. According to a report in the ESECS study, attacking WTO main air bases would be a very critical and urgent mission for NATO. If a WTO air attack is pending and the Eastern air forces are to be hit on the ground or prevented from using the runways, a quick-reaction force of long-range ballistic missiles is necessary and an attack 'must take place within 15 or 30 minutes to be successful.'[89] Thus, the military requirements for a real-time response could, in fact, come close to a call for pre-emption.

Furthermore, if deep-strike weapons are connected to a strategy of counter offensives with ground forces crisis stability would be even more seriously affected. The US Army's ALB doctrine with its emphasis on winning a military engagement envisages offensive-orientated 'blitzkrieg'-type, deep manoeuvre warfare into enemy territory including a terrain-seizing capability.[90] If such a concept were adopted by NATO, it could mean that the Western alliance might carry out deep attacks with ground forces up to 150 km (at corps level) into East European territory. If such a capability for ground counter-offensives is at the same time coupled with offensive political motives of threatening the cohesion of the Eastern alliance, then the situation becomes very precarious. Huntington, for instance, has argued in favour of a conventional retaliatory strategy for NATO. This implies a conventional offensive capability in order to 'put at potential risk the system of controls over Eastern Europe that the Soviets have developed over thirty years and which they consider critical to their own security'.[91] Any step towards a posture with such an offensive orientation and capability threatens to undermine the relative stability of the European status quo and poses enormous risks for crisis stability. If the WTO were to perceive a military conflict as inevitable, there would in the face of such a threat be great incentives to strike first before NATO's armed forces were fully deployed.

On the othe hand, according to statements by General Rogers, offensive deep attack with ground forces as envisaged in the extended battlefield concept of the ALB doctrine is not compatible with the current NATO strategy and doctrine.[92] Thus, the ALB doctrine reflects war goals and operational concepts which are in apparent contradiction to NATO's declaratory defensive strategy.[93] In spite of this, the US Army Europe is trained and equipped according to its ALB doctrine. There are also indications that elements of the ALB doctrine are finding applications among the armed forces of NATO. According to one report, the Northern Army Group (NORTHAG) has introduced a revised operational concept so as to be able to fight a mobile defence battle as a unified command. This concept will, according to the British NORTHAG commander, allow NATO forces to seize the initiative, to launch a surprise counter-attack, and ultimately to win: 'The quicker we get on the offensive, the better'.[94] Corresponding phrases can be found in the ALB doctrine. The transition to a mobile defence as an alternative to the war of attrition concept means that the Western Alliance would have to extend its capability for offensive operations. This would also present a potentially unstable situation in a crisis. As Mearsheimer remarks:

> If NATO's forces were capable of conducting a mobile defense, those same forces would surely have a significant offensive capacity . . . a mobile defense is actually a very offensive-minded strategy. Its principal ingredient is the counterstroke, which is an offensive tactic. An army that could satisfy the demands of a mobile defense would be well suited for conducting offensive operations.[95]

In view of these possibilities, the author concludes that the new US/NATO offensive-orientated war-fighting doctrines and the associated projected weapons for deep strike embody potential grave consequences for crisis stability. In order to be 'successful', FOFA or ALB demand very quick employment decisions which could very well border on pre-emption. This could imply that a military conflict in Europe might rapidly be geographically extended and the escalation process thereby accelerated and intensified. This, in turn, would require further, quasi-automatic military responses by NATO. Given the incentives or coercion for pre-emptive considerations, which both the WTO and NATO would then weigh up in a crisis situation, the new

Western war-fighting doctrines threaten to accelerate events which would run counter to NATO's interest in gaining room for political actions and decisions.

Political control and crisis decision-making on the automated battlefield

In Western democracies, the principle that politics has a clear-cut precedence over the military is undisputed. However, the primacy of politics is called into question by the new US/NATO war-fighting doctrines and related efforts to introduce the automated battlefield. Some general problems in the interaction between politics and the military have laid the foundation for this development.

Especially with regard to conventional weaponry, politicians are scarcely able either to measure military demands against political criteria or—if necessary—to refuse these demands. Usually, the military puts forward demands originating from their doctrines which have been drawn up without due regard for political purposes. Politics threatens to become dependent on military premises if military operational concepts and weapons are approved in advance without evaluation of their manifold consequences. This problem is aggravated by the fact that military organizations show an inclination towards offensive operations and a general resistance to intervention by the civilian political leadership in any operational planning.

Military organizations, like all large organizations, tend to seek autonomy from outside influences. Thus, in peace time, civilians are seldom exposed to the intricacies of military planning, and in wartime, when civilian intervention in the details of military policy is much more likely, soldiers often interpret policymakers' injunctions in ways that allow them maximum operational discretion. . . . Historically, offensive action, which requires complicated, detailed, expert planning, has been a way for the militaries to evade civilian control. Under current conditions, this pattern suggests that . . . policymakers may have the least influence over the most escalation operations.[96]

Thus, the realm of politics could become part of a dynamic process with its own momentum resulting from purely military considerations—a process from which it has to withdraw in order to maintain its primacy. If—because of the shift towards the offensive—the military doctrine of NATO does not take into consideration the political purpose of the alliance in its operational planning (which according to its stated declaratory policy is defensive), then the primacy of politics threatens to become lost in the thinking of the military.

The pace of new armaments technologies and their doctrinal implications pose the danger of outrunning the ability of the political authorities to understand and influence these technologies and to maintain political control. If politicians were to be constrained by current doctrines and military planning, then it is conceivable that during an intense crisis a war might rapidly break out and moreover nuclear weapons would be resorted to early on in the conflict. In order to be 'successful', new war-fighting doctrines like FOFA or ALB demand a quasi-automatic military decision at the earliest possible point of time. The reduced decision time inherent in these new concepts would leave the political authorities little or almost no time for appropriate consideration and response. The real-time requirements of the new doctrines, and the associated necessity for a high degree of computerized automation of the C^3I processes, could force the policy-makers into new operational modes which endanger the political control process. Possibly all political barriers to escalation would have to be removed in favour of a kind of automatic mechanism if FOFA or ALB are to achieve their military objective. The complexity of the battlefield of the future, the increasing proliferation of highly sophisticated weaponry and the need for real-time decision-making could force the political authorities to assign an increasingly larger proportion of their decision-making authority to computers in automated battlefield management systems.

V. Conclusion

Many current developments in armaments technology are clearly heading towards the automation of warfare on the tactical battlefield, as discussed above. The trend is, as far as possible, to remove human influence and control from the automated war machine since human beings are not able to process the immense mass of battlefield data or to handle the control of the increasingly complex weapon systems. But how would the transition from peace to war look in an automated battlefield management system? Who would decide that an attack reported by sensors, for example tank forces moving towards the border, is unequivocal? Automated battlefield management systems rest on complex technology. As experience has so far shown it is very questionable whether this complex technology would be reliable in practice. On the contrary, such technology could increase the probability of a military conflict, for example, through signal processing errors. Thus, highly automated battlefield management systems, which imply the removal of humans from the real-time decision-making loop, could drastically reduce the possibility of recognizing and correcting technical errors during actual military engagement.

Furthermore, how would the safeguarding of civilian political control over military matters provided for constitutionally actually look and work on the future automated tactical battlefield? What would be the consequences for the primacy of politics under the presupposition of automated decisions? A former US Deputy Director of Defense, Research and Engineering illustrated this situation in 1976: '. . . It is of no use to give (the President) a room full of status boards and say, "Here it is, boss, make a decision". It has to be boiled down to a scale—for example green, yellow, or red—and he can decide by how far the needle moves, what he should do'.[97] However, are the decision-makers capable and informed enough to control the movement of the needle and to choose the proper colour? Human limitations in speed and accuracy when assessing situations and taking decisions are already regarded as a central problem in the increasingly complex, *conventionally* automated systems. This problem is to be solved in future command and control systems by the introduction of AI embodying 'intelligent' computers with the ability to infer and to deduce. Yet, this development could mean that the possibilities for influence by the higher decision-making levels will probably be mostly limited to simple decisions, including the decision to substitute one automatic response by another. Such automated decisions would then produce a cascading sequence of automatic interactions among the many sections of the complex military force structure controlled by 'intelligent' machines—with no other human interference.

The ultimate version of the automated tactical battlefield is one in which advanced 'intelligent' computers, exploiting various sensors and using different weaponry in a closed loop, would form *one* single enormous weapon system in which humans would be totally removed from the loop. Once one side started a given set of actions, the other side would automatically initiate counter-moves. The automated war machines for the tactical battlefield would begin to work and they would stimulate each other beyond the control of human beings. The spirit of such a process was captured by the Austrian novelist Robert Musil: 'The machine was there; and because it was there, it had to work, and once it was running, it began to accelerate'.[98] The progressive application of AI technology for military purposes could in the end very well lead to such fatalistic machines.

Notes and references

[1] For a good description see Dickson, P., *The Electronic Battlefield* (Indiana University Press: Bloomington & London, 1976).

[2] Quoted by Dickson (note 1), appendix A, pp. 220–21.

[3] The FOFA and ALB concepts are now covered by a wide body of literature. For more details and discussion of their doctrinal differences and similarities, see for example Berg, P. and Herolf, G., 'Deep strike: new technologies for conventional interdiction', in SIPRI, *World Armaments and Disarmament: SIPRI Yearbook 1984* (Taylor & Francis: London, 1984), pp. 291–318; Sutton, B. D. *et al.*, 'Deep attack concepts and the defence of Central Europe', *Survival*, vol. 2 (Mar.–Apr. 1984), pp. 50–70; Gessert, R. A., 'The AirLand Battle and NATO's new doctrinal debate', *RUSI Journal for Defence Studies*, vol. 129, no. 2 (June 1984), pp. 52–60; for a more critical political account see Coolsaet, R., 'NATO strategy: under different influences', *ADIU Report*, vol. 6, no 6 (Dec. 1984), pp. 4–8; Plesch, D. T., 'AirLand Battle & NATO's military posture', *ADIU Report*, vol. 7, no. 2 (Mar.–Apr. 1985), pp. 7–11; and Nikutta, R., 'Eine offensive Kriegsführungsdoktrin für das Schlachtfeld Europa', Working Paper AP 09 (M&P), Berghof Foundation for Conflict Research, Berlin, Jan. 1984.

[4] For a more detailed discussion see Berg and Herolf (note 3), pp. 295–97, and Herolf, G., 'Emerging technology', in SIPRI, *World Armaments and Disarmament: SIPRI Yearbook 1986* (Oxford University Press: Oxford, 1986), pp. 193–208.

[5] Britton, J. D., *Computers in the Army: Applications and Implications in the Year 2000*, US Army College, Strategic Studies Institute, Carlisle Barracks, Pennsylvania, 1983, pp. 8–9.

[6] US Army, *Equipping the United States Army*, Statement to the Congress on the FY 1986 Army RDTE and Procurement Appropriation (US Government Printing Office: Washington, DC, 1985), p. X/7–8.

[7] US Army (note 6), pp. XII-5.

[8] US Army (note 6), pp. XII-3/4; Schultz, J. B., 'PLRS, PHJ to improve tactical battlefield operations', *Defense Electronics*, vol. 16, no. 1 (1984), pp. 60–71.

[9] Wagner, L. C., 'Soldier-first attitude has "new prominence"', *Army*, vol. 35, no. 10 (1985), p. 250.

[10] NATO ACCS', *Internationale Wehrrevue*, vol. 19, no. 9 (1986), pp. 1359–60.

[11] Ogorkiewicz, R. M., 'Automatisierte unbemannte Kampfpanzer und Roboterfahrzeuge', *Internationale Wehrrevue*, vol. 19, no. 9 (1986), pp. 1283–90.

[12] Lavenson, J., 'Human factors considerations for C^3I', *Army Research, Development & Acquisition Magazine*, vol. 24, no. 1 (Jan.–Feb. 1983), pp. 14–16; Ehrenreich, S. and Moses, F. L., 'Get the message? The problems of abbreviations and battlefield automated systems', *Army Research, Development & Acquisition Magazine*, vol. 22 (Jan.–Feb. 1981), pp. 14–16.

[13] Gerencser, M. and Smetek, R., 'Artificial intelligence on the battlefield', *Military Technology*, vol. 8, no. 6 (1984), p. 86.

[14] Note 13.

[15] Cooper, R. S., Prepared Statement to the Committee on Armed Services, *Hearings on Department of Defense Authorization for Fiscal Year 1985*, Part 4, Research, Development, Test and Evaluation, US Congress, House of Representatives, 98th Congress, Second Session (US Government Printing Office: Washington, DC, 1985).

[16] This summary is especially indebted to an overview study by Bankes, S. C., *Future Military Applications for Knowledge Engineering*, RAND Note N-2102-1-AF, Santa Monica, CA, 1985; and to Andriole, S. J., 'Artifical intelligence comes of age', *National Defense*, vol. 69, no. 403 (Dec. 1984), pp. 43–52; and Schultz, J. B., 'Weapons that think', *Defense Electronics*, vol. 15, no. 1 (1983), pp. 74–80.

[17] Taylor, E. C., 'Artificial intelligence in the Air-Land Battle', *Astronautics & Aeronautics*, July/Aug. 1983, p. 55.

[18] Callero, M. *et al*, *TATR: A Prototype Expert System for Tactical Air Targeting*, RAND Report R-3096-ARPA, Santa Monica, CA, 1984.

[19] See for example Clark, F. G., 'The commander and battlefield automation', *Military Review*, vol. 64, no. 5 (1984), pp. 67–71; Williams, J. D., 'Leadership in the computer age', *US Naval Institute Proceedings*, vol. 109/11/969 (Nov. 1983), pp. 140–42; Mac Laren, jr, W. G., 'The force multiplier, C^3 in NATO', *Military Technology*, no. 3 (1984), p. 24; and Rose, K. H., 'Why artificial intelligence won't work', *Military Review*, vol. 66, no. 12 (1986), pp. 57–63.

[20] Interview with Gary Martins, in Schultz (note 16), p. 78.

[21] Taylor (note 17), pp. 55–56.

[22] Taylor (note 17), p. 56

[23] US Army (note 6), p. II-2.

[24] 'Project BETA—what is it', *Army Research, Development & Acquisition Magazine*, vol. 13, no. 2 (Mar.–Apr. 1978), p. 13; Schultz (note 8), p. 59; and Schultz (note 16), p. 79.

[25] Weiss, A. H., 'An order of battle advisor', *Signal*, vol. 41, no. 3 (1986), pp. 91–95.

[26] Taylor (note 17), p. 58.

[27] Taylor (note 17), p. 59.

[28] Canby, S. L., 'New conventional force technology and the NATO-Warsaw Pact balance', in IISS, *New Technology and Western Security Policy*, Part II, *Adelphi Papers*, no. 198 (IISS: London, 1985), pp. 17–18.

[29] US Army (note 6), p. II-2/3.

[30] Bankes (note 16), p. 32.

[31] Bankes (note 16), p. 32.

[32] Bankes (note 16), p. 20.

[33] Klahr, P. *et al*, *TWIRL: Tactical Warfare in the ROSS Language*, RAND Report R-3158-AAF, Oct. 1984.

[34] Marsh, R. T., 'A preview of the technology revolution', *Air Force Magazine*, vol. 67, no. 8, pp. 42–49.

[35] Bankes (note 16), pp. 19, 21.

[36] Cooper, R. L., Statement to the Committee on Armed Services, *Hearings on Department of Defense Authorization for Appropriations for Fiscal Year 1986*, Part 4, Research, Development, Test, and Evaluation, Title II, US Congress, House of Representatives, 99th Congress, First Session (US Government Printing Office: Washington, DC, 1985), p. 651; and Morishige, R. I. and Retelle, J., 'Air combat and artificial intelligence', *Air Force Magazine*, vol. 68, no. 10 (1985), pp. 91–93.

[37] Clark (note 19).

[38] Gutzmann, L. E. and Hogge, S. M., 'Navy applications of artificial intelligence', *Army Research, Development & Acquisition Magazine*, vol. 26, no. 6 (Nov.–Dec. 1985), p. 5; Walters, B., 'Processing images like the human eye', *Jane's Defence Weekly*, vol. 6, no. 20 (1986), p. 1247.

[39] Bankes (note 16), p. 34.

⁴⁰ Richardson, K., 'Tactical missile technology for the 1990's', *Military Technology*, vol. 10, no. 10 (1986); and Runge, P., 'Intelligente munition', *Jahrbuch für Wehrtechnik*, Vol. 16 (Bernard & Graefe Verlag: Koblenz, 1986), pp. 202–11.

⁴¹ Schultz (note 16), p. 75.

⁴² Bankes (note 16), pp. 34–35.

⁴³ Pan, K. C., 'Applications of robotics and artificial intelligence to armament', *Army Research, Development & Acquisition Magazine*, vo. 24, no. 5 (Sep.–Oct. 1983), p. 15.

⁴⁴ Bankes (note 16), p. 36.

⁴⁵ Cooper (note 15), pp. 925–26.

⁴⁶ Ulsamer, E., 'The next computer generation', *Air Force Magazine*, vol. 68, no. 6 (1985), pp. 87–93.

⁴⁷ Lynch, R. and McGee, M. R., 'Military applications of artificial intelligence and robotics', *Military Review*, vol. 66, no. 12 (1986), pp. 50–56.

⁴⁸ Pan (note 43), p. 17.

⁴⁹ Bankes (note 16), pp. 46–48.

⁵⁰ Ludvigson, E. C., 'Light forces reshaping modernization program', *Army*, vol. 34, no. 10 (1984), p. 326.

⁵¹ Gerencser and Smetek (note 13), p. 86.

⁵² Bankes (note 16), p. 15.

⁵³ Canby (note 28), p. 8.

⁵⁴ Allen, R. and Psotka, J., 'Artificial intelligence for the executives', *Army Research, Development & Acquisition Magazine*, vol. 26, no. 6 (Nov.–Dec. 1985), p. 2.

⁵⁵ Bankes (note 16), p. 8; Frith, S., 'Software development planning', *Army Research, Development & Acquisition Magazine*, vol. 26, no. 2 (Mar.–Apr. 1985), pp. 15–16; and Canan, J. W. 'The software crisis', *Air Force Magazine*, vol. 69, no. 5, pp. 46–52.

⁵⁶ Gerencser and Smetek (note 13), p. 92.

⁵⁷ Bankes (note 16), p. 8; and Kimmel, S., 'Very High Speed Integrated Circuits for Army systems', *Army Research, Development & Acquisition Magazine*, vol. 25, no. 6 (Nov.–Dec. 1984), pp. 11–13.

⁵⁸ For an overview see Leitenberg, M., 'Background information on tactical nuclear weapons', in SIPRI, *Tactical Nuclear Weapons: European Perspectives* (Taylor & Francis: London, 1978), pp. 3–136.

⁵⁹ Bracken, P., 'The political command and control of nuclear forces', *Defense Analysis*, vol. 2, no. 1 (1986), p. 17.

⁶⁰ Bracken, P., *The Command and Control of Nuclear Forces* (Yale University Press: New Haven and London, 1983), p. 138.

⁶¹ Sinnreich, R. H., 'NATO's doctrinal dilemma', in Pranger, R. J. and Labrie, R. P. (eds), *Nuclear Strategy and National Security* (Points of View: Washington, DC, 1977), p. 306; see also Burt, R., 'The hidden nuclear crisis in the Atlantic Alliance', in Yost, D. S. (ed.), *NATO's Strategic Options* (Pergamon Press: New York, 1981), pp. 46–59.

⁶² Wiles, R. I. *et al*, ORI Inc., *A Net Assessment of Tactical Nuclear Doctrine for the Integrated Battlefield*, Report prepared for the Director of Defense Nuclear Agency, NTIS AD A100504.

⁶³ Van Cleave, W. R. and Cohen, S. T., *Tactical Nuclear Weapons: An Examination of the Issues* (Macdonald and Jane's: New York, 1978), p. 24.

⁶⁴ Frei, D., *Risks of Unintentional War*, UNIDIR (UN Publications: Geneva, 1982), pp. 103 and 71.

⁶⁵ For details see Hanne, W. G., 'The integrated battlefield', *Military Review*, vol. 62, no. 6 (1982), pp. 34–44; and Starry, D. A., 'Extending the battlefield', *Military Review*, vol. 61, no. 3, pp. 31–50.

⁶⁶ Romjue, J. L., *From Active Defence to AirLand Battle: The Development of Army Doctrine 1973–1982*, TRADOC Historical Monograph Series, Fort Monroe, VA, 1984, p. 37.

⁶⁷ Din, A. M., 'Strategic computing', in SIPRI, *World Armaments and Disarmament: SIPRI Yearbook 1986* (Oxford University Press: Oxford, 1986), p. 187.

⁶⁸ Saxer, R. K., Director of the Defense Nuclear Agency, Prepared Statement to the Committee on Armed Services, *Hearings on Department of Defense Authorization of Appropriations for Fiscal Year 1985*, Part 4, Research, Development, Test, and Evaluation, Title II, US Congress, House of Representatives, 98th Congress, Second Session (US Government Printing Office: Washington, DC, 1984), p. 987.

⁶⁹ Note 68.

[70] Saxer, R. K., Prepared Statement to the Committee on Armed Services, *Hearings on Department of Defense Authorization of Appropriations for Fiscal Year 1986*, Part 4, US Congress, House of Representatives, 99th Congress, First Session (US Government Printing Office: Washington, DC, 1985), pp. 278–79; and Wiles (note 62).

[71] Ball, D., 'Can nuclear war be controlled?, *Adelphi Papers*, no. 169 (IISS: London, 1981).

[72] Romjue (note 66), p. 37.

[73] US Army, *Field Manual FM 100-5*, (Department of the Army: Washington, DC, 1976), p. 10-9.

[74] Romjue (note 66), p. 37; Wiles (note 62), pp. 120–21; and Tomhave (note 62), p. 19.

[75] Pincus, W., 'Army would like advance authority to use A-weapons', *Washington Post*, 21 July 1982.

[76] Rogers, B. W., interview in *Wehrtechnik*, no. 1 (1985), p. 14.

[77] Rogers, B. W., 'Die langfristige Planungsrichtlinie FOFA: Behauptungen und Tatsachen', *Amerika Dienst* (US Information Service), 19 Dec. 1984, p. 10.

[78] Goure, D. and Cooper, J. R., 'Conventional deep strike: a critical look', *Comparative Strategy*, vol. 4, no. 3 (1984), p. 228.

[79] Posen, B. R., 'Inadvertent nuclear war?', *International Security*, vol. 7, no. 2 (Fall 1982), p. 29.

[80] Posen (note 79), p. 35.

[81] Posen (note 79), p. 28.

[82] Charles, D., 'Two paths to stability in Europe', *Journal of the Federation of American Scientists*, FAS Public Interest Report, vol. 37, no. 5 (May 1984), p. 7; Berg and Herolf (note 3), p. 309–10; and Klare, M. T., 'Conventional arms, military doctrine and nuclear war: the vanishing firebreak', *Thought*, vol. 59, no. 232 (Mar. 1984), p. 63.

[83] This aspect is discussed in more detail by Posen (note 79).

[84] See Posen (note 79) for a detailed discussion.

[85] Frei (note 64), p. 10.

[86] Goure and Cooper (note 78), p. 227; for a detailed discussion of their offensive aspects see also Lübkemeier, E., 'AirLand Battle und Rogers-Plan', *Die Neue Gesellschaft*, no. 4 (1984); Nikutta, R., 'Kommentar und Replik: "AirLand Battle"-Konzept', *Aus Politik und Zeitgeschichte*, no. B7-8 (16 Feb. 1985), pp. 30–33; and Afheldt, H., *Defensive Verteidigung* (Rowohlt: Reinbeck, 1983), pp. 24–42.

[87] Cotter, D. R., 'Potential future roles for conventional and nuclear forces in defense of Western Europe', *Strengthening Conventional Deterrence in Europe: Report of the European Security Study (ESECS)* (Macmillan: London & Basingstoke, 1983), pp. 236–44.

[88] Charles (note 82), p. 7.

[89] Cotter (note 87), p. 223.

[90] See, for example, Woodmansee, J. W., 'Blitzkrieg and the AirLand Battle', *Military Review*, vol. 64, no. 8 (1984), pp. 22–39.

[91] Huntington, S. P., 'Conventional deterrence and conventional retaliation in Europe', *International Security*, vol. 8, no. 3 (winter 1983), p. 42; Some West German military officers argue for a similar offensive defence based on forward motion, see for example Farwick, D., 'Zur Diskussion der NATO-Strategie. Dynamische Vorwärtsverteidigung statt statischer Vorneverteidigung', *Österreichische Militärische Zeitschrift*, vol. 21, no. 2 (Mar.–Apr. 1983), pp. 117–20.

[92] Rogers (note 76), p. 1985.

[93] For a more detailed discussion see Nikutta (note 86).

[94] O'Dwyer-Russell, S., 'NORTHAG concept "aims to win" ', *Jane's Defence Weekly*, vol. 6, no. 3 (26 July 1986), pp. 116–17.

[95] Mearsheimer, J. J., 'Maneuver, mobile defense, and the NATO central front', *International Security*, vol. 6, no. 3 (winter 1981), p. 120.

[96] Posen (note 79), p. 32.

[97] Quoted by Aldridge, R. C., *The Counterforce Syndrome: A Guide to US Nuclear Weapons and Strategic Doctrine* (Institute for Policy Studies: Washington, DC, 1987), p. 64.

[98] Quoted by Ford, D., *The Button. The Pentagon's Strategic Command and Control System* (Simon and Schuster: New York, 1985), p. 52.

Chapter 7. Software and systems issues in strategic defence[1]

HERBERT LIN

I. Strategic defence

Strategic defence missions include defence against ballistic missiles—ballistic missile defence (BMD); air-breathing threats such as bombers and cruise missiles—air defence (AD); and weapons to destroy satellites—anti-satellite weapons (ASAT). Of these, BMD is the most demanding, owing to the large volume of information that such a system must handle (thousands of missiles, tens of thousands of warheads, hundreds of thousands or millions of decoys), and the short time in which it must process this information (15–30 minutes).[2]

It is likely that any future large-scale BMD system will be integrated with systems that perform other missions. For example, BMD operations would have to be co-ordinated with AD and ASAT operations to avoid interference. The hardware used for one mission of strategic defence may be used to support another mission; if current procurement practices are followed in the future, the hardware may be used to perform missions for which it was not initially designed.[3] For instance, a space-based BMD system would need to protect itself; self-protection facilities would strongly resemble facilities for ASAT. Moreover, the close technological relationship between ASAT and BMD missions would mean that a system capable of comprehensive BMD would also be capable of highly effective and prompt satellite destruction. For these reasons, it can be expected that a future BMD system would also be given an ASAT role simply on the grounds of cost-effectiveness. A BMD system could also have some ability to assist in air defence.

US strategic defences would also be integrated with other US strategic assets, namely the early-warning system and the strategic offensive forces. Such integration would pose problems and challenges in addition to those that would be faced by a purely defensive system. For example, it is likely that a BMD would have options to support a strategic offensive attack that might be executed to back up a collapsing conventional defence of Western Europe.[4] In this role, a BMD could be required to perform two functions. In particular, the USA has allocated for NATO use some 400 submarine-launched ballistic missile (SLBM) warheads. Since a Soviet ground-based missile defence could impede the penetration of these warheads, a US BMD system could have the job of suppressing Soviet missile defences. Since Soviet interceptors could not be launched until incoming US missiles were detected, the operation of the US

BMD in defence-suppression mode would have to be co-ordinated with the launch of US offensive forces. The second function of a US BMD would be to destroy Soviet missiles fired at the USA in retaliation for the US SLBM strike. To perform this task, it might be necessary to distinguish between Soviet and US SLBMs, both of which could be launched from large and overlapping areas of the ocean.

The current US strategic posture is structured primarily around offensive elements, as is the Soviet strategic posture to a lesser degree. Consequently, strategic defences will be added incrementally on to an existing array of strategic hardware and procedures. As such, strategic defences will be constrained to a considerable extent by elements of this existing system, resulting in design compromises.

At the time of writing, no specific architecture for a system intended to perform comprehensive BMD has emerged as a clear favourite. Nevertheless, a few general observations can be made. For example, a defence consisting of several layers is likely to be more effective than one consisting of a single layer. However, a multi-layered defence would in general cost more, as well as involve a higher degree of technical risk than a single layer.

Space-based lasers in low earth orbit for boost-phase interception of ballistic missiles have been de-emphasized in current research programmes. Instead, current architectural sketches suggest that short-term defences will emphasize non-nuclear ground-launched interceptor missiles for terminal and late mid-course interception, and sensors using space-based and air-based infra-red detectors and ground-based radar. Architectures for the longer term would augment short-term systems with space-based platforms launching interceptors for mid-course interception, interactive discrimination using neutral particle beams for sorting decoys from re-entry vehicles (RVs) in mid-course, and ground-based free electron lasers using space-based relay mirrors for boost-phase interception.

A system intended putatively for BMD may or may not be capable of performing other missions, depending on the hardware ultimately involved. For example, powerful anti-missile lasers operating at an appropriate wavelength may theoretically have significant anti-aircraft or even ground-attack capabilities. Moreover, choices for specific operating parameters such as wavelength are not normally mandated by Congress, and indeed are not usually subject to Congressional oversight or review.

II. Software and systems

One challenge for BMD is to develop individual components (e.g. weapons and sensors). A second challenge is to integrate these components into an effective and functional military system. Computers will control the operation of virtually all aspects of that system. They will control the function of individual weapons (e.g., the guidance of a missile, the aiming of a laser), process data from the sensors to identify, recognize, and track potential targets and finally, they will perform battle management (i.e., monitoring of the threat and the progress of the battle, and the co-ordination of the weapons available to meet the threat). Indeed computers will, in most cases, 'replace human decision-making'.[5] As a result, issues of computer programming *per se* cannot be considered in isolation from issues of system design and architecture. In this paper, the term 'software' is used broadly to include the manner in which the many hardware elements of a system are bound together in a particular architecture, the analysis and specification of tasks that the computer-controlled system must perform, the computer programs that direct the system to perform these tasks at the appropriate times, the documentation that allows human beings to understand the limits of its capabilities and to change its operation through re-programming, and the things that human beings must do in order that the system perform correctly. All discussion of software for BMD must assume a particular configuration of weapons, sensors and system architecture. The relevant issue is the extent to which the behaviour of the system, controlled by that software and architecture, can be understood and predicted.[6] A highly idealized model of the software development process can be characterized as a process in which developers:

(*a*) analyse system requirements;
(*b*) generate functional specifications of system performance;
(*c*) construct a design that will meet these requirements;
(*d*) implement this design in an actual set of computer programs;
(*e*) integrate these programs with each other and other system elements;
(*f*) test the system in operational environments;
(*g*) train personnel to operate the system properly; and
(*h*) maintain and repair the system if it behaves in unanticipated or inadequate ways.

In practice, these activities are not conducted separately; rather, they overlap and interact with each other to a significant degree. For example, the results of operational testing may alter system specifications and integration difficulties may result in changes to design. Some of the salient issues that arise in the development of software for BMD are addressed below; for additional discussion, the reader is referred to other sources.[7]

Analysis and design

The analysis of requirements, the generation of specifications, and the construction of a design are the most difficult aspects of a project, since these aspects depend primarily on human judgement. While developments in software engineering over the past 15 years do make it easier to articulate requirements and specifications that are less ambiguous and more consistent, they cannot (and should not) carry the burden of human decision-making. In addition, since the time-scale that characterizes the BMD problem is too short to allow human intervention, systems analysts must anticipate the characteristics of all environments in which the defence would have to function, and the appropriate response in these environments.

As the operating environments become more complex, the difficulty of determining the functional requirements of a system to perform in those environments increases, as more human decisions must be made correctly. The problem is made even harder when systems analysts have no empirical experience of fighting a nuclear war, or of battle management involving thousands of ballistic missile warheads, that will allow them to determine the environments that the system must be able to handle or what system behaviour might constitute 'failure'.

As one example of these difficulties, the 'rules of engagement' should be considered—the circumstances under which a missile should be attacked. For instance, should a US BMD system be designed with the capability to shoot down US missiles? If it is, it may accidentally shoot down US missiles launched in retaliation to a Soviet attack. If it is not, it may be unable to shoot down a US missile launched by accident.

Even the determination that a hostile missile has been launched is not easy to specify clearly. A simple specification might stipulate that the launch of a missile would be signalled upon the detection of an electromagnetic signal of 10-micron peak emission wavelength and a net radiated power of 100 kW or more. A determination of hostile intent might be made if the ground point of origin of the signal was located within Soviet missile fields.

These criteria are not unreasonable, but they are incomplete in important ways. For example, a booster exhaust from a space-shuttle launch might also emit an electromagnetic signal of 10-micron peak-emission wavelength and a net radiated power of 100 kW or more. A signal originating from within a Soviet missile field may be hostile (or it may be a missile test), but hostile missiles may also be mobile and thus their launching points may not be located within current missile fields.

Another example concerns the specifications required for the software that processes the streams of raw data emerging from sensors. Mistaken assumptions about the nature of the environment (e.g., the effects of nuclear bursts or the countermeasures used against the sensors), or about the signatures of re-entry vehicles and decoys, could reduce dramatically the efficiency of the defence. For example, an incorrect assumption about the time for 'nuclear red-out' in infra-red sensors to subside could lead to reduced detection efficiency and thus to the loss of some warheads in mid-course against the background of space. An incorrect assumption about the signature of re-entry vehicles could cause many warheads to be mistaken for decoys. An incorrect assumption about the signature of decoys could cause decoys to be mistakenly identified as targets, causing computer overloads or the expenditure of resources against non-threatening objects.

In addition to difficulties related to the specific problem of ballistic missile defence, other difficulties characteristic of large-scale military projects also occur. For example, the complexity of requirements analysis and system design usually leads to specifications and designs that are incomplete. The details of these specifications are then left to those responsible for implementation, that is, the programmers. These individuals generally lack the global knowledge to complete these specifications sensibly, and so may 'fill in' the details in a way that is not appropriate from a total system perspective. Worse yet, these individuals often do not realize explicitly that these specifications are incomplete, and may fill in the details subconsciously, resulting in different (and sometimes conflicting) interpretations of the requirements. Finally, military systems must function in an environment in which an enemy has powerful incentives to invalidate the assumptions made by the designers and analysts of the original system. The history of programming illustrates that it is difficult enough to anticipate all of the situations that arise in a co-operative environment. It can only be more difficult to anticipate the actions of a malevolent opponent who goes out of his way to create new situations. Indeed, an enemy directly controls many of the requirements the system must meet by choosing the actual circumstances of any attack—timing, tactics, hardware. While the laws of physics and economics do place important fundamental constraints on the range of possible attack scenarios, many top-level decisions regarding attack scenarios will depend on intelligence estimates, which are fundamentally softer, less reliable, and subject to greater circumstantial variability.

Evidently, it is impossible to anticipate, and therefore to discuss, a situation that cannot be anticipated. But the general problem remains. The issue is not 'How likely is it that the system will contain a particular error or omission?' but rather 'How likely is it that the system will have any one of uncounted potentially fatal and undetected errors and omissions?'. The primary matters of concérn are those that remain unimagined—the 'unknown unknowns'. These difficulties arise from the deficiencies in human thought processes, and are therefore hard to examine in the absence of tangible demonstrations. Because computer programs are essentially mental constructs expressed as lines of computer code, it is easy to postpone the confrontation between ideas and reality to a time much later than is possible for those who build physical objects. The result is that the clarity perceived by analysts and designers is not matched by the actual existence of clarity, and errors of omission and commission occur.[8] Moreover, the discovery of unanticipated errors during an attack will leave no time for humans to program or even suggest a new and more appropriate system response.

To summarize, the ability of BMD-system designers to specify requirements in adequate detail is limited by two problems: it is hard to anticipate all the circumstances that can arise and it is often difficult to know the appropriate actions for the circumstances that can be anticipated. Thus, it is easy to understand the response of the former Director of the Defence Advanced Research Projects Agency (DARPA), Robert Cooper, to a question about the problems pacing the development of ballistic missile defence. Cooper noted that: 'we have no way of understanding or dealing with the problems of battle management in a ballistic missile attack ranging upward of many thousands of launches in a short period of time'.[9]

In the three years since Cooper made this statement, no empirical experience has accumulated to alter this assessment.

Reliability

A BMD system for comprehensive defence must be thoroughly reliable. Analytical tools to assess software reliability can only assess the extent to which an actual program conforms to its specifications; they cannot determine the extent to which the specifications are themselves correct. In addition, it is very difficult to understand how a complex computer program will perform, since existing mathematical tools are essentially incapable of predicting the behaviour of systems that are not continuous. As a result, even small extrapolations from known system behaviour cannot be trusted.

In particular, a computer program is a model of reality, in which 'important' aspects of a complicated situation have been retained, and other 'unimportant' aspects discarded in order to facilitate analysis. Thus, one must make decisions about what is or is not sufficiently important to include, and about how to take account of structures about whose 'internals' we are ignorant. This set of decisions (a 'decision space') cannot in general be analysed by any mathematics known today. In particular, this decision space is not continuous, and the omission of just one particular aspect could render the model entirely useless for practical purposes. As an example, a decision to omit as 'too far-fetched' the possibility that space mines (satellites that trail space-based BMD platforms and that explode if themselves attacked or on command) could pose a threat to a BMD system would result in a system architecture that would be highly vulnerable if the Soviet Union chose to threaten a US BMD in this way. Only a model that took into account the possibility of space mines could predict this vulnerability.

In addition, computer programs are quantized—the smallest change is one bit of information—and 'small' differences between programs can result in very large differences in their output. As a result, the notion of a model that is 'approximately correct', in the sense of a model that is just 'a little bit' different from an 'absolutely correct' model, is not useful for software. For example, in June 1985, the Strategic Defense Initiative Organization (SDIO) conducted an experiment to bounce a laser beam off a mirror mounted on the space shuttle. The test was a complete failure, because the altitude of the ground-based laser was entered as 11 023 miles rather than the correct 11 023 feet. The error was small while the consequences were large. Yet, no mathematical tools could have predicted the outcome of this experiment.[10]

In the absence of analytical tools to determine reliability, software engineers attempt to 'design in' reliability. For example, rather than designing systems that operate without error, they design 'fault-tolerant' systems that can operate adequately even when errors occur. While such techniques improve reliability, they can themselves lead to failure. The use of redundancy is a good example. Redundancy is used to ensure that when one component fails, the function it performed can be taken over by other components. However, redundancy requires the installation of additional machinery to sense failure and to direct a 'spare' component to assume the function of a component that has failed. The result is additional system complexity, which may lead to a loss of analytical transparency (making it harder for people to understand), and more opportunities for error. For example, the first attempt to launch the space shuttle failed as the result of the shuttle designer's decision to provide redundancy in its computers—a synchronization error between these machines aborted the launch. These potential liabilities must be weighed against the increase in component-level reliability provided by redundancy.

In general, the only way to assess the reliability of a system is to subject it to the stress of actual operating environments. It is necessary to perform empirical testing under a variety of circumstances large enough that one can begin to have confidence that the system will perform as expected even when unlikely events occur. Empirical testing reveals the existence of errors that would otherwise remain hidden. For most large programs, it is essentially impossible to test the correctness of all possible sequences of program execution, since the number of these sequences grows roughly exponentially with the size of the program. Of course, a software system for ballistic missile defence cannot be subjected even once to a large-scale empirical test under realistic conditions short of nuclear war.

The extreme difficulty of large-scale empirical testing for BMD is recognized, but proponents have suggested that a combination of small-scale testing and the use of computer simulators to mimic large-scale threats could substitute for actual system-wide testing. For example, the Eastport panel (chartered by the SDIO to examine the feasibility of software for ballistic missile defence) suggested that ballistic missile defence should be designed around 'loosely co-ordinated' architectures that are not dependent on global communications; instead, components would share threat information only with nearby neighbours.[11] With such an architecture, weapons would be allocated to targets without the benefit of information obtained earlier in the target's trajectory. The computational benefit would be a significant reduction in the necessary co-ordination between components; the only cost acknowledged by the Eastport panel would be a certain fraction of wasted 'shots' caused by non-optimal allocation. Furthermore, the panel alleged that this design would reduce the possibility that failures due to inconsistencies in the threat data base or programming errors would propagate throughout the system. This would result in the ability to infer total system performance on the basis of individual component testing and simulation testing, thereby eliminating the need for full-scale system tests.

This type of system architecture has desirable characteristics, but it has difficulties as well. For example, component independence is a desirable trait, but total independence is difficult to achieve. The targets are shared among platforms, in the sense that the trajectory characteristics of the various targets are the same for all platforms. The load placed on a particular layer depends on the effectiveness of previous layers in destroying re-entry vehicles and identifying decoys; the collapse of a boost-phase defence could hopelessly overload subsequent mid-course and terminal defence layers. The top-level specifications (e.g., the rules of engagement) are 'shared', in the sense that they would be identical for each platform; thus, a specification error would most likely be common to each platform. The actual programs that control the operation of each type of weapon will be similar if not identical. Finally, it is not necessarily true that defensive technologies that are more capable of independent operation are also more capable from a physical point of view. For instance, a space-based weapon platform might require sensors of its own to function independently; these sensors could reduce the mass available for greater battle endurance. Therefore, system designers may be forced to trade independence of operation in return for a highly capable weapon. Finally, it is generally accepted that loosely-coupled systems are harder to develop, debug, and maintain than either centralized or completely independent ones.

In general, while successful small-scale tests are necessary for a system to work, they are far from sufficient.[12] The underlying reason is that difficulties of 'system integration' (ensuring that all pieces of the system work well together even when operating near the limits of their capabilities) tend to appear only when the system is tested near its limits.[13] Indeed, prudent system developers generally 'margin test' their systems at levels somewhat in excess of their rated load.

The use of computer-driven simulators provides some large-scale testing opportunities not otherwise available. Simulators do provide valuable information and confidence in system operation beyond that achievable by small-scale testing alone. Nevertheless, simulation testing is limited, because it presumes that the simulation developers can anticipate in accurate detail the natural and man-made environment that will result from the set of all possible attacks. Such an assumption is not warranted. As one example, a strongly nuclear environment presents simulation developers with important uncertainties about the physical processes underlying this environment; these uncertainties cannot be resolved with confidence in the absence of multi-burst atmospheric testing of nuclear weapons, and the Limited Test Ban Treaty prohibits all atmospheric nuclear tests.

In addition, simulators are themselves software systems that are as much as 10 times larger than the computer programs they help to exercise. As such, they may themselves contain undetected bugs. For example, during the development of the radar-guided Phoenix air-to-air missile, the system was unable to track simulated targets even when they were flying in a straight line and not using electronic countermeasures; it was later discovered that the radar simulation was faulty.[14]

System evolution

The idealized model in which a threat is defined and a system developed to address that threat is overly simplistic. Not only will the system be changed after reaching its initial operating capability (due to errors that are uncovered in the course of routine testing), it will also be changed as the Soviet Union responds or as US hardware improves. In other words, as changes in the threat (new tactics, hardware, and so on) arise, or as more capable hardware becomes available, the system will be re-programmed to accommodate the new possibilities inherent in these changes.

Problems can arise when existing software is changed to accommodate new hardware. For example, a defensive system involving 150 laser battle stations in space might well include different models of the 'same' battle station, which (hypothetically) could have different response functions for mirror settling and thus require slightly different software controlling the mirror-moving mechanisms. It is not difficult to imagine that an error correction intended for one model might inadvertently be made instead on software for another. This updating error might not be noticed until actual operational use. Other problems could arise in the event that during a time of crisis, US intelligence reports suggest that the Soviet Union could attack in a way that has not been considered in planning. Under these circumstances, software developers could be under extreme time pressure to implement changes quickly, with the result that new software would probably be installed without the benefit of testing as extensive as that undergone by the initial version of the software, resulting in a higher probability of failure.

To summarize, the growth of large-scale software is well described in the following quotation:

When a program grows in power by an evolution of partially understood patches and fixes, the programmer begins to lose track of internal details, loses his ability to predict what will happen, begins to hope instead of know, and watches the results as though the program were an individual whose range of behaviour is uncertain. This is already true in some big programs, [and] it will soon be much more acute . . . [Programs] will be developed and modified by several programmers, each [acting] independently . . . The program will grow in effectiveness, but no one of the programmers will understand it all. (Of course, this won't always be successful—the interactions might make it get worse, and no one might be able to fix it again!).[15]

III. Failure modes

Types of failure

A system can fail in one of two ways: it may fail to perform the appropriate actions when the external circumstances dictate that it should, or it may perform certain actions when the circumstances dictate that it should not.

Thus, the strategic offensive system may fail to launch when properly ordered to do so, or it may launch when it has not been ordered to do so. A BMD system may fail to operate when hostile missiles are launched, or it may operate even when no hostile missiles have been launched. These failure modes have different consequences depending on the system involved. A nuclear offensive system that operates under the wrong circumstances is likely to result in nuclear devastation and war. A BMD that operates under the wrong circumstances may have direct results that are less consequential by comparison. By contrast, a failure to operate properly under the appropriate circumstances has major consequences for a BMD, but possibly less major ones in the case of an offensive system.

This is not to say that the operation of a BMD system at an inappropriate time would necessarily be harmless. While a BMD based on non-nuclear interceptor missiles might be physically unable to destroy targets other than ballistic missiles or satellites, its operation under inappropriate circumstances may trigger an unanticipated sequence of unlikely events at an inappropriate time that might have disastrous consequences. In other words, while the direct consequences of inappropriate BMD operation might be relatively small, the indirect consequences could be large.

For example, on 9 November 1979, a false alert of missile attack was generated by NORAD when a test tape was mistakenly run. Imagine a similar occurrence in a strategic context involving a BMD system relying on nuclear 'pop-up' X-ray laser interceptors based on submarines relatively close to Soviet shores. Under these circumstances, a false alert could transmit a mistaken warning to these launch platforms, resulting in the near-simultaneous launch of many interceptors from below the ocean surface. The proximity of launch platforms to the Soviet Union—necessary for popped-up X-ray lasers to intercept missiles during their boost-phase—would also reduce the amount of time between the launch of submarine-launched ballistic missiles (SLBMs) and their impact. Thus, the Soviet Union—probably unable to distinguish between interceptors with a defensive purpose and SLBMs with an offensive purpose—would be forced to make a very rapid decision as to whether or not it were under attack. Since SLBMs in the near future will be capable of destroying missile silos, the Soviet Union would have to consider the possibility of launching its ICBMs before impact. The necessity of considering a 'launch-on-warning' would not increase stability.

Sabotage

Countermeasures can take the form of different attack timing, tactics and hardware. However, it is not unreasonable to expect the Soviet Union to attempt to infiltrate the software development process as well. Consider two forms of computer sabotage: 'trap doors' and 'time bombs'.

A 'trap door' is that part of a program that enables a knowledgeable person to bypass the security in the system and perform unauthorized actions. Space-based BMD satellites will most likely be equipped with remote reprogramming capabilities. A 'trap door' written by a programmer with Soviet sympathies might enable the Soviet Union to transmit an unauthorized software change to a US BMD satellite, possibly allowing the Soviets to take control of the satellite, or—more likely—rendering the software on board useless.[16] A 'time bomb' is the part of a program that performs as expected most of the time, but can misbehave unexpectedly when triggered by very specific circumstances. For example, a time bomb might cause a laser to misfire at all times after, say, 14:32:21 on 19 July 2013.

The usual response to this problem—security clearances—can make software development a far more difficult process. High-level security is based on compartmentalization. But when errors due to system-integration problems occur, they occur precisely because of unexpected interactions between different parts of the system. Reducing the communication between people responsible for different parts of the system can inhibit the information transfers necessary to identify these errors.

The man-machine interface

Computers control other machinery as part of a system. The system is in principle controlled by human beings, whose role it is to make decisions and exercise judgements that should not be delegated to the computer. The human interface to the computer (the so-called 'man-machine interface') is an enormously important influence on system behaviour.

In particular, the end users of a system are usually not the software developers of the system, and they will have minimal knowledge of the assumptions embedded in the software. Consequently, difficulties or questions that arise during system operation cannot be addressed immediately. Small decisions made by a programming team years earlier may have consequences that no current user (or even programmer) will understand.

The new system will accept raw data from the sensor inputs, and process it to make the appropriate inferences. High-level decision-makers will most likely receive processed information, rather than the raw data; indeed, it may be very difficult to even retrieve the raw data in any reasonable time.[17] If the system receives ambiguous or uncertain information, the processing algorithms will yield results of uncertain significance. While this uncertainty could be apparent to users, they would not know how to resolve it without detailed knowledge of the underlying processing algorithms. Moreover, they would be highly unlikely to have access to anyone who did have such knowledge. The result may be that users become locked into interpretations of uncertain data that are based on choices outside their control.

These comments suggest that under some circumstances, a 'man-in-the-loop' will not be able to prevent system error; the personnel involved might not have the expertise to understand the situation as it unfolds.

IV. Control

Escalation control is a process by which crises are managed so that they do not turn into wars, or small wars into large wars. Within the context of military forces emphasizing nuclear offensive capabilities, the mechanism for achieving escalation control is the use of small fractions of one's warheads on appropriate targets. For forces emphasizing defensive capabilities, escalation control cannot be based on partial use—it is very difficult to imagine scenarios in which it would be desirable for missile defences to perform less than perfectly. Rather, if conflict escalates, it is likely to involve a failure in some scenario other than one involving an actual missile attack. The following scenarios illustrate this:

1. A US BMD system is deployed with many space assets in low earth orbit. Those assets are protected by an automatic self-defence that destroys any unknown object within two kilometres. During a time of crisis, a Soviet radar satellite is manoeuvred with the intention of monitoring a US carrier battle group; however, by chance, one manoeuvre places the Soviet satellite into an orbit that would take it within this keep-out zone. The Soviet satellite is destroyed accidentally, as the consequence of standing US rules of engagement. An automatic Soviet defence of its non-BMD satellites returns fire to the attacking platform, causing the USA to react as though its BMD were under attack.

2. A BMD is designed with a rule of engagement in which only the terminal and mid-course layers of a defence are activated during peacetime. Boost-phase defence would be activated only during crisis or war. Assume that the ground-based facilities for boost-phase intercept take several minutes to reach full readiness. In the early stages of a crisis, uncertain tactical intelligence reaches BMD command headquarters that suggest the Soviet Union is preparing to launch a large missile strike against the USA. The USA orders its boost-phase intercept layer to its highest readiness as a precaution. However, the Soviet Union interprets the US preparations as evidence that the USA is preparing to launch a first strike against them and decides to pre-empt, rather than wait for US defences to be brought to full capacity.

3. One assumption underpinning US nuclear strategy is that the execution of a selective nuclear option with small numbers of weapons is likely to be less escalatory than an all-out strike. The USSR could deploy a limited defence against ballistic missiles to frustrate attacks based on this assumption. However, the USA could decide that a particular military objective would be worth the allocation of many weapons, and carry out a many-weapon strike for a limited objective on the assumption that the Soviet defence would destroy a considerable fraction of the attacking warheads. The failure of Soviet BMD software could ensure that this attack, intended to have limited effects on the ground, will instead have large collateral damage, thereby resulting in pressures for escalation.

4. A test tape is accidentally loaded and mistakenly run, improperly activating the ASAT capabilities of the BMD system. There is no time for human controllers to halt the near-instantaneous destruction of many Soviet satellites. The Soviet Union fears the beginning of a US first strike.

Despite these scenarios, the effects on escalation control of a highly automated network for strategic defence including ASAT, AD and BMD are hard to predict. Indeed, it is also possible to imagine ways in which the deployment of a BMD could make escalation control easier. For example, BMD based in space would require many orbiting platforms. In some BMD system architectures, these platforms would be designed to communicate with one another; these communication facilities could also be designed to conduct space-to-ground communications. Thus, under some circumstances, these channels could provide an additional route by which the National Command Authority could communicate with strategic offensive forces in the field. A second scenario is one in which a US ballistic missile defence could destroy a US strategic ballistic missile launched in error, thereby averting accidental escalation.

V. Artificial intelligence

Some analysts believe that artificial intelligence offers an alternative approach to software development. Artificial intelligence approaches that are alleged to hold such promise are those of expert systems and automatic programming.

Expert systems are computer systems designed to embody the knowledge of human experts as computer programs, albeit in restricted domains. The term 'expert system' describes 'the codification of any process that people use to reason, plan, or make decisions as a set of computer rules, . . . [involving] a detailed description of the precise thought processes used.'[18]

To date, expert system research has focused on well-defined areas such as biochemistry and internal medicine. However, many believe that expert systems have matured to the point that they can be used over a much broader set of applications. Indeed, one specific focus of the DARPA Strategic Computing Initiative is the implementation of an expert system to assist in battle management for carrier battle groups, with software spin-offs from this demonstration project that would strongly support battle management for missile defence.[19]

This point of view neglects the fact that human expertise (the ultimate source of knowledge to be included in any expert system) is based on human experience. No one has 'expert knowledge based on experience' concerning the battle management task for missile defence. Moreover, even ostensible experts are often confused when they encounter radically unfamiliar situations, and make judgements based on past experience that may be entirely incorrect within the context of the new situation. Worse still, reasonable people can disagree on what an 'appropriate' response is. This potential difference of opinion leads to difficulty in deciding if some particular behaviour of a program is desirable or undesirable. As programmers put it, 'Is that behavior a bug or a feature?'.

Certain types of expert system have their own characteristic weaknesses. One very common type of expert system is typically based on a set of 'if-then' rules, called productions. In its basic form, a production rule states that 'If X occurs, then take action Y.' A very fundamental problem then arises. System designers must anticipate the set of 'X' that can occur. What if something happens that is not anticipated in the specified set of 'X'? (This leaves aside the difficulties of determining the set 'Y' of reasonable actions.) The most common result of such circumstances is that nothing happens.

A second type of expert system employs a representation of knowledge about the problem called a 'frame'. A frame might represent a generic table as an object with four legs, a flat surface on top, and made of wood. Frame-based systems make extensive use of default values, so that your kitchen table might be specified as a generic table, but with three legs. More generally, if a particular value is not specified for the situation at hand, the default value characteristic of such situations is used instead. While the use of default values does mirror the way in which humans often make 'common-sense assumptions', the inappropriate use of default values—especially if used without explicit recognition—can lead to confusion. For example, a frame-based system for object identification might categorize any object seen in space as a re-entry vehicle unless it is specifically identified as a decoy. A decoy might be indicated if the object had any of 10 sets of particular characteristics. If the Soviet Union deploys a decoy that does not conform to any of these 10 sets, the frame-based system would falsely identify many decoys as real targets.

Finally, many expert systems use informal reasoning procedures whose validity cannot be proved or disproved using standard tools of formal logic. (For example, the converse of a true statement is not necessarily true (or false), but an expert system might use it anyway as a 'rule-of-thumb'.) Consequently, developers of expert systems are strongly motivated to program by 'giving advice' to the system that 'seems right' under the circumstances; the ability to 'program' in this manner is often alleged to be a virtue of expert systems. However, this advice may lead to local improvements to performance without revealing deeply embedded conceptual contradictions on a system-wide scale that might cause catastrophic failure at some (unknowable) point in the future.[20]

A second allegedly promising technology is automatic programming: the use of programs to write other (error-free) executable programs. For example, Fletcher states that battle management software for BMD 'will require the development of techniques for automated software development—in essence, a computer that can write another computer program.'[21] Major Simon Worden (now Lieutenant Colonel) of the Strategic Defence Initiative Office has stated that 'A human programmer can't do this. We're going to be developing new artificial intelligence (AI) systems to write the software. Of course, you have to debug any program. That would have to be AI too.'[22] More generally, automatic programming systems are said to enable people with little or no programming expertise to program computers, and to enable professional programmers to generate code (translate ideas into computer language) with much greater ease than if they had to do so without computer assistance. Indeed, some in the military software community predict that a person will be able to 'talk directly to the machine in natural language, make a few sketches, wave his arms a bit, answer a few questions, say yes or no a few times, and have the program tested and documented in the morning.'[23]

This optimism is not justified. In particular, automatic programming systems do not (indeed, should not) relieve the human burden of decision-making imposed by the analysis and design process, which generates about two-thirds of all errors in computer software.[24] Rather, the primary functions of automatic programming systems are to facilitate the implementation of design specifications into actual code and the modification of currently existing code. Even the most optimistic assessments of automatic programming acknowledge that the decision-making aspects of the system-design process cannot be automated to any significant degree.[25]

VI. Deterrence versus defence; confidence versus reliability

The discussion to this point has focused on a BMD system whose task is to protect the US population against large-scale nuclear missile attack. For this purpose, near-perfect performance and reliability would be required. However, in order to derive the benefits of such a defence, the USA would have to be confident that the defence would actually perform as designed if placed into use. Thus, it is important to distinguish between reliability and confidence. Reliability is a characteristic of the system in use, describing the extent to which the actual behaviour of the system (as controlled by software) matches its expected behaviour; however, reliability cannot be specified precisely in the absence of operational experience.[26] By contrast, confidence is a characteristic of human beings who must make decisions about the use of the system. Confidence describes the extent to which human beings believe that the system will exhibit its desired behaviour; thus, confidence can be high, but misplaced.

These distinctions have profound significance. An untested system may or may not perform with very high reliability when it is first placed into operation. But decision-makers could not reasonably have high confidence in its reliable operation, for reasons discussed previously. Without confidence in the operation of a BMD system, the USA would be forced to take account of the possibility that the BMD might not protect the nation. Thus, it would still require the ability to retaliate against the Soviet Union to deter an attack. A similar argument answers various proponents of strategic defence who assert that for purposes such as the defence of military assets, lower levels of performance and reliability can be tolerated. Proponents suggest that even these lower levels would create uncertainty that would complicate the task of planning an attack on the USA, thereby reducing Soviet incentives to strike and contributing to deterrence. However, if the USA did not maintain retaliatory capabilities, the Soviet Union would have little to lose by attacking the USA. Thus, in this case, the USA would also require the ability to retaliate against the Soviet Union.

Finally, while it is true that BMD would increase the uncertainty of Soviet planners, other methods would also do the same; the cost-effectiveness of BMD in creating uncertainty must be compared to that of other methods. For example, the current Administration has itself noted that a comprehensive test ban (CTB) would reduce the reliability of nuclear weapons; a CTB would be cheaper to implement than most forms of BMD, and would certainly present lower technical risk.

VII. Conclusion

This survey of software issues for strategic defence suggests that despite the high degree of automation inherent in any plausible defence against ballistic missiles, it is unlikely that computer malfunctions in a 'stand-alone' defensive system would by themselves lead to accidental nuclear war. While accidental activation as the result of computer error can lead to catastrophe in the case of the nuclear offensive forces, no comparable scenario is plausible in the case of a defensive system. However, system errors in a defence that is integrated with offensive forces may have indirect effects, in which interactions between the different components of each nation's strategic forces (both offensive and defensive) lead to unexpected and undesirable consequences. Artificial-intelligence approaches, new techniques for software development and architectural choices are not likely to alter this conclusion, because the limits of confidence in the reliable operation of a complex system are set by the limits of the human imagination and its inability to anticipate all possible circumstances.

Notes and references

[1] The assistance of David Benson, William Durch, Bob Estell, Paul Dietz, Jon Jacky, Jim Horning, Dave Redell, Bill Wessomer, Richard Hilliard, Peter Mikes, Michael Scott, Peter Denning, Steve Litvintchouk, Pamela Lasky, Richard Kulawiec, Scott Preece, Len Gaska and Karen Weickart is gratefully acknowledged. They are of course not responsible for any errors or misstatements remaining in this chapter.

[2] BMD can be deployed on many scales for many purposes. BMD has been variously justified as a comprehensive defence of populations, a limited defence of selected military assets, a defence against missiles launched by accident or by terrorists, and a defence against the few missiles that an adversary might secretly deploy in violation of an arms agreement. Unless otherwise stated, the BMD discussed in this paper is assumed to be for comprehensive population defence. A general description of how BMD would proceed can be found in any of hundreds of articles and books with the words 'ballistic missile defence' or 'Star Wars' in their titles.

[3] It would of course be possible to refrain from adding additional missions, but it is natural for planners faced with tight budgets to expect 'double duty' from existing systems. BMD system developers would be wise to resist all such demands (especially since the task of developing reliable

and effective missile defences is difficult enough without other distractions), though the history of weapon procurement does not suggest that their resistance would be fruitful.

[4] Indeed, for many years the USA has had options for launching pre-emptive strikes against the Soviet Union contained within the Single Integrated Operating Plan controlling the strategic forces; the use of a US BMD system to support these options would be wholly consistent. See Ball, D., 'US strategic forces: how would they be used?', *International Security*, vol. 7, no. 3 (winter 1982/83), p. 38.

[5] Fletcher, J., 'The technologies for ballistic missile defense', *Issues in Science and Technology*, Fall 1984, p. 21.

[6] The rationale for this approach is persuasively argued by Leveson, N., *Software Safety: Why, What, and How*, Technical Report 86-04, University of California at Irvine, Feb. 1986. Many of the systems issues raised by Leveson (and discussed below) are relevant to all engineering projects as well.

[7] Lin, H., *Software for Ballistic Missile Defense*, Report C-85/1 (Center for International Studies, MIT: Cambridge, MA, June 1985), and 'The development of software for ballistic missile defense, *Scientific American*, Dec. 1985, pp. 46–53; Parnas, D., 'Software aspects of strategic defence systems', *Communications of the ACM*, vol. 28, no. 12 (Dec. 1985), pp. 1326–35. A good discussion of systems issues related to a ballistic missile defence, which provides inspiration for some of the points raised in this paper, is Zraket, C., 'Strategic defense: a systems perspective', in *Weapons in Space, Vol. I: Concepts and Technologies, Daedalus*, vol. 114, no. 2 (spring 1985), pp. 109–126.

[8] Brooks, F. P., *The Mythical Man-Month* (Addison-Wesley: Reading, MA, 1978), p. 15.

[9] Cooper, R., 'Strategic and theater nuclear forces', Hearings before the Senate Committee on Armed Services, *Department of Defence Authorization for Appropriations for FY 1984* (US Government Printing Office: Washington, DC, 1983), Part 5, p. 2892.

[10] Built-in 'sanity checks' to check that data conform to expected ranges would have helped in this instance. Nevertheless, sanity checks have two major disadvantages. The first is that sanity checks will pass data that are plausible though wrong. The second is that they will block data that are implausible but correct.

[11] Eastport Study Group, 'Summer study 1985: a report to the Director of the SDI Organization', Dec. 1985.

[12] Dr Donald Latham, current Deputy Under Secretary of Defence for C^3I, has pointed out that 'a one-on-one test is going to be meaningless; [the problem is] how do we conduct a many-on-many test?' Quoted in James Canan, 'Fast Track for C^3I', *Air Force Magazine*, July 1984, page 43.

[13] A recent study performed at the computer facilities of the Stanford Linear Accelerator Center notes that the rate of system errors increases with system workload, and that such errors are by far most common when the system is heavily loaded.

[14] Jim Horning, Digital Equipment Corporation, personal communication.

[15] Minsky, M., 'Why programming is a good medium for expressing poorly understood and sloppily formulated ideas', in Krauper, M. and Seitz, P., *Design and Planning 2* (Hastings House: New York, 1967), p. 121. A personal communication with Professor Minsky confirms that he still holds this view.

[16] The idea that the Soviet Union could take control of a US satellite is not entirely outrageous; the author has heard from a reliable source that the ground signals to the Soviet ocean reconnaissance radar satellite are unencrypted, and that during a crisis the USA would simply turn the Soviet satellite off.

[17] NASA ground control personnel had no indication that the Challenger was in danger before it exploded in January 1986, even though subsequent analysis of data transmitted from the shuttle but not displayed to controllers indicated that something was wrong.

[18] Defence Advanced Projects Research Agency, *Strategic Computing*, 28 Oct. 1983, p. 7.

[19] Note 18, pp. 27–29.

[20] These comments should not be taken to suggest that expert systems are never appropriate. In domains in which extensive empirical experience provides the basis for expertise and for tasks in which humans can intervene in the event of inappropriate system behaviour, experts systems hold considerable promise; neither of these conditions is met by the BMD problem.

[21] Fletcher (note 5), p. 25.

[22] Quoted in *Washington Post*, 3 Mar. 1985, p. 7.

[23] Compare Everett, R., 'Command, Control and Communications', *The Bridge*, National Academy of Engineering, vol. 11, no. 2 (summer 1981).

[24] Compare Wegner, P. and Grafton, R., 'Perspective on software engineering', *Naval Research Reviews*, no. 4 (1982), p. 18.

[25] Compare Green, C. *et al.*, *Report on a Knowledge-Based Software Assistant*, Rome Air Development Center, RADC-TR-83-195, Aug. 1983, pp. 25–32.

[26] The term 'reliability' is used in its everyday usage, not in its technical usage of mean time to failure or number of remaining bugs.

Chapter 8. Artificial intelligence and disarmament

GENNADY B. KOCHETKOV, VLADIMIR P. AVERCHEV AND
VIKTOR M. SERGEEV

I. Introduction

The rapid development of artificial intelligence in recent years has given rise to a wide literature on its possible applications. Particular attention has been paid to potential uses since the launching of a number of large-scale national and international technological programmes such as the 'Fifth Generation Computer' (Japan), 'ESPRIT' (EEC) and the 'Strategic Computing Initiative' (USA). Moreover, the member countries of the Council of Mutual Economic Assistance have adopted the 'Complex Program of Scientific and Technological Progress of Member-Countries of CMEA up to the Year 2000'.[1]

The general developments in information technology also affect the military domain. New technology, and especially one of its 'offspring', AI systems, is increasingly becoming the augmenter and amplifier of human intellect. This will eventually lead to qualitative changes in the core military–strategic and political doctrines. At present these far-reaching consequences are largely neglected and the attention of researchers and experts focuses mainly on tactical and technical parameters and the cost-efficiency of the projected weapon systems. Such technocratic bias may be partly explained by the phenomenon of 'two cultures' described by C. P. Snow, which is inherent in the military domain as well as in many other spheres of human endeavour.[2]

During periods of technological breakthrough and creation of radically new weapons, polarization of the world perceptions of the members of these two cultures increases. One pole is represented by members of the 'engineering culture', who are interested mainly in the technical capabilities of the new systems. The opposite pole is composed of military–political experts who try to single out how the new military hardware will affect the attainment of the pre-established military–political objectives.

It is extremely important to realize that the new technology cannot be reduced to just a new means of attaining the unalterable objectives. This new information technology creates a whole new world with its own laws and principles. The implementation of advanced military systems transforms the entire military–political situation. Some of the 'residual' effects of the implementation process may sometimes have much more impact on the structure of the situation than the immediate and intended goals of military planners. In this new world, the very process of formation of political and strategic objectives is changing. A whole set of new difficulties arises in the domain of arms control—and the rapid progress of the new information technology in the military field makes the task of achieving international consent on matters of arms control especially urgent. As General Secretary Gorbachev stated, 'There must be no delays, otherwise weapons of such a degree of sophistication will be created that it will be impossible to come to an agreement about their control altogether.'[3]

Any major innovation in the military field should, therefore, be introduced with great caution. This point was stressed by one of the founders of Soviet military science, M. V. Frunze.[4] This warning is especially urgent today since the fate of our world greatly depends on the ability to control effectively the military applications originating in the current technological revolution. At present, mankind faces a choice—to build a safer world or to enter a new round of the arms race based in large part on the military application of new information technology. A course of events which would result in the adoption of new 'intelligent' weapon systems and eventually in a less safe and less stable world must be avoided. Meanwhile it appears that awareness of the nature of the 'new world' is developing too slowly.

Emphasis on the military–strategic aspects of the large-scale introduction of information technology into weapon systems was augmented with the announcement of the US Strategic Defense Initiative (SDI). However, the discussion has revolved mostly around issues of feasibility and reliability of the proposed weapons; at the same time urgent and fundamental problems such as the impact of SDI on the stability of the world military–political situation were largely ignored. This chapter is an attempt to outline, in general terms, the new world situation which is being created with the advent of new information technology in the military domain.

II. Technological progress and the military–political balance

One of the central concepts of modern military–political analysis is the notion of the balance of power. That is why we begin with a discussion of the possible impact of the large-scale introduction of modern information technology on this balance.

The power-balance concept is rooted in the ideas of political balance introduced for the first time by Machiavelli and Lord Bolingbroke.[5] This general idea was later transformed into the concept of power balance, with its stress on the analysis of the components of military power.

The Napoleonic wars led to profound changes in the concept and exercise of military force. The advent of an army of 'armed people' instead of a relatively small professional army sharply increased the importance of a country's mobilization capabilities. This rapid growth of the numerical stength of armies stimulated an intensive search for more effective technical means of destruction. The arms race had begun.

Technological advances coupled with the growth of the military industry have gradually become the major source of change in the armed forces. As Engels[6] noted, 'Nothing depends on economic conditions more than the Army and Navy. Arms, manpower, organization, tactics, and strategy depend in the first place on the attained stage of production and on the means of communication at any given moment'. Thus, the invention of the steam turbine made it possible to build heavy high-speed battleships; and the invention of the internal combustion engine and the emergence of tanks led to the decline and eventual elimination of cavalry. It is important to note that each major technological breakthrough in the military field has radically changed the composition and relative imporance of different components of the power equation among the major world military powers, and triggered a new round of the arms race under the revised military balance. A striking example of this kind of race was the British–German rivalry in the area of navy building from the end of the nineteenth century to the beginning of the twentieth century.

In the second half of the twentieth century, the sharp acceleration of technological progress has resulted in revolutions in weapon systems as well as in the very notion of military power. Mutual intersections of the lines of technological development have become more and more frequent, leading to quantum changes in the capabilities of armaments with often unpredictable consequences for the global military–political situation. Simultaneously, efforts to put the arms race under political control are increasing. Numerous attempts are made to create some kind of structure of international bilateral and multilateral agreements which may function as 'a cage' for the armaments and hold them within secure boundaries. Moreover, it becomes increasingly clear that the radical resolution of arms control problems presupposes an equally radical reformation of political thinking and the current international military–political structures.

The emergence of nuclear weapons, and especially the development of the 'strategic triad' (intercontinental ballistic missiles, nuclear submarines and strategic bombers), has led to a revolution in the notion of military power. The intersection of the two lines of technological development within the framework of established structures of military power, namely the increasing yield of explosives and the invention of missiles, had unforeseen consequences: fundamentally new weapon systems emerged, characterized by the combination of the whole range of highly lethal effects, such as blast waves, flash radiation, radioactive contamination and the electromagnetic pulse.[7]

In the First and Second World Wars, the main targets of combat operations were the enemy's masses of infantry, air forces and navy. The depth of strike was limited by the attack capabilities of the air forces and the final strategic mission accomplished through the number of local successes achieved in combat actions, battles and operations.

With the advent of nuclear weapons the limits of combat operations have become sharply widened. Strategic nuclear forces make it possible to conduct operations throughout the territory of the enemy coalition and to accomplish the main missions of the armed conflict in a short period of time.[8] The almost total absence of efficient protection against nuclear attack made this weapon an absolute one.

These developments resulted in the total reorganization of military power. The strategic nuclear triad occupied the dominant place and the state's military might became determined mainly by the possession of nuclear weapons and the sophistication of the means of delivery. Simultaneously, the research and technological potential in the military sphere became a major factor of the strategic posture. These are examples of the fundamental changes in the capabilities of conflicting parties in the middle of the twentieth century which have taken place under the modern technological revolution.

The world in which military–political decisions are made and implemented has changed. In particular, the initial period of a future war acquires a greater importance, since it may turn out to be a very decisive period for the outcome of the entire arms conflict.[9]

As the time span for critical military–political decisions is sharply reduced, the probability of losing control over weapon systems increases.[10] This new situation creates powerful incentives to develop systems for data processing and command, control and communications, that would provide reliable data bases for decision-making under stringent time constraints. In other words, the inherent logic of development of modern weapon systems inevitably requires the most advanced information technologies to be encompassed in this process.

III. New information technology

It is the view of many experts that large-scale introduction of these edge-cutting technologies will result in drastic changes in the possible modes of war-fighting. They will be characterized by an increased scale and mobility of battlefield operations, engagement of various arms, abruptness of changes in the combat situations, fierce fighting for seizing and holding terrains, and so on.[11] In this way the army is transformed into a complex socio-technical system which can accomplish its missions only through the organic merge of human and machine capabilities. Accordingly, the complexity of military command and control tasks is immensely exacerbated.

New opportunities for making a quantum leap in the effectiveness of military command and control are created with the rapid progress of information and computing technologies. The development of integrated computerized management systems enables flexibility, steadiness, reliability and continuity of war-fighting command and control. This, in turn, sharply reduces the time and effort required to organize combat actions and improves co-ordination of various arms.[12]

Thus, the intelligence factor is becoming a new major component in the power equation. It can no longer be considered negligible by comparison with nuclear weapons and other factors of military might. Hence, the need arises for a conceptual approach to the assessment of this intelligence factor in strategic equations for different countries and regions. This must involve an assessment of the effects of introducing artificial intelligence into particular components of computerized management and military decision-making. The functions of computerized command and control systems are illustrated in table 8.1.

Table 8.1. The functions of computerized command and control systems

System	Function
1. Information sensing 2. Pattern recognition	Reconnaissance and early-warning systems
3. Data interpretation 4. Data integration	Expert system
5. Situation forecast	Human being
6. Interference and assessments 7. Generation of action alternatives	Expert system
8. Decision-making	Human being Expert system
9. Command and control 10. Monitoring	Automated battlefield command and control system

The applications of AI in this cycle were recently limited to the fields of reconnaissance, early warning and missile guidance. With the development of more advanced expert systems, the role of AI in war-fighting command and control is increasing. Such critical functions as data interpretation and integration, and the generation and assessment of alternatives are gradually being computerized. Man retains some rather poorly structured functions in situation forecasting and (partial) choice of action.

These general principles of AI applications can be traced at different levels of military–political decision-making, strategic, operation and combat command and control. The introduction of 'intelligent technologies' in reconnaissance, communications, data processing, monitoring and other systems enables the time allocated for making critical decisions under the tight constraints of crisis or armed conflict situations to be 'stretched'.

However, it is extremely important to be aware of the limitations of the present state-of-the-art artificial intelligence systems. The essence of modern AI is computer presentation and processing of knowledge.[13] Expert systems designed to assess situations and to generate action alternatives are based on the methodology of knowledge engineering.[14] The logical structure of such systems is mainly built on the simple productions of the 'if-then' kind. A comprehensive description of the 'natural inference' procedures is yet to be developed, in spite of several attempts to simulate human reasoning.[15]

The next generation of expert systems may be characterized by the ability to generate new knowledge.[16] While AI systems are based on rather primitive rules of production, one can hardly expect serious improvements to result from their application to the assessment of strategic or military–political situations.

That is why many experimental military AI applications do not meet operational requirements in terms of effectiveness, reliability, and so on.

None the less, dangerous trends are apparent in the international military–political situation. Some Western politicians, especially those with close ties to the military–industrial complex, pretend that all technical difficulties are already resolved. They try to sell to the general public military–political doctrines based on potential, rather than actual, capabilities of the 'intelligent' military technologies and in this way try to mobilize political support for the development of AI-based weapon systems. This gap between technology and politics is one of the major sources of destabilization in the current international scene.

The central unresolved problems of AI application concern the role of the human factor in the 'intellectualized' weapon systems.[17] Modern expert systems represent 'snap-shots' of the skills of real specialists, developed through long practical experience. In the military domain there are principle differences among expert systems for automated control of the 'smart weapons', for battlefield command and control, and those for higher-level decision making.

In the case of strategic command and control, the practical experience required to develop adequate expert systems is almost totally absent since the knowledge acquired through war simulation games or through testing the modern weapon systems is far from sufficient in real-life war-fighting situations. That is why the practical value of expert systems for strategic-level missions is highly doubtful.

There are certain reasons for believing that the provocation of local conflicts is used more and more as a testing ground for the new 'smart weapons'. According to the Western press, such goals were pursued by the USA and Israel in a number of local conflict situations (for example the air-attack on Libya and the Bekaa Valley incident). Those 'experiments' demonstrate the destabilizing effect of the accumulation of 'smart weapons' and point to the dangers of relying on AI techniques in large-scale projects of the SDI type; such projects may create a false impression of reliability and security.

The above-mentioned danger stems first of all from the fact that expert systems built into weapons will inevitably face two types of contingency:

1. Some critical parameters of the situation may overstep the limits stipulated by the design of command systems (i.e., low-altitude launching paths or increased velocity of missiles, etc.);

2. The situation may change in directions not covered by the scenarios implemented in the expert system.

The unquestionable advantage of the human factor in decision-making systems lies in the human ability to single out and assess negative aspects of complex situations. In the case of automated command and control this advantage is lost. Hence, in certain situations, AI-based systems could behave in conflict with human values.

It is already clear that artificial intelligence will operate in close working contact with the 'natural intelligence' of men for any level of automation of the war-fighting command and control. This situation presents complex problems concerning the interaction between artificial-intelligence systems and human beings and the formation of a combined intellect of the man-machine system. These problems are far from being resolved. But it seems evident that in such systems with 'hybrid intellect'[18] it is impossible to predetermine every possible situation, simply because human behaviour is not totally predictable.[19] Human behaviour is determined not only by the kind of data to hand but also by the system of values and moral norms of society. Of course, we are far from suggesting the development of a normative system for controlling machine behaviour like the famous 'Laws of Robotics' invented by Isaac Asimov. But since we are dealing with hybrid systems that involve men, the behaviour of such systems in unforeseen situations must be determined by human values, norms, and rules.[20]

Social values and norms cannot be invented and introduced arbitrarily; they are the result of the long evolution of mankind. Information technologies introduce a new channel for the transfer of social norms—through the program product. It is now possible to imprint these norms in a computer memory and its knowledge bases, and the choice of the values and norms to be 'implanted' into the command and control element of a weapon system through artificial intelligence cannot be considered a strictly technical task. Since it has a decisive impact on world peace and stability, this is a humanitarian and political choice.

We want to stress yet another critical issue that is often neglected. It is well known that one of the central problems of AI system design concerns extraction pattern recognition and interpretation of data about the external world.[21] The majority of such interpretation models produce a large number of errors. In the case of highly specialized AI systems, those errors may be within acceptable limits so that the system will fulfill prescribed tasks satisfactorily. But the problem of the possible mode of behaviour of a complex system, which consists of many 'imperfect' expert systems, is quite another matter. We are far from being inclined to revive the fear of machines that may rebel against their creators, though this theme was very popular in the classic literature on cybernetics.[22] But the plan to deploy an SDI system whose command and control elements would inevitably be equipped with complex expert systems puts the issue of human control back on the agenda. Thus the AI applications in the military domain face two types of serious constraint. The first is related to the technical shortcomings of present artificial intelligence systems. It may be overcome in the foreseeable future through research in the field of cognitive modelling. The second set of constraints stems from the socio-technical nature of the war-fighting command and control systems and cannot be overcome by technical means. Instead it presupposes the creation of an adequate system of political control on the national and, especially, the international level. That is why it seems important to discuss some possible military–political consequences of the large-scale introduction of AI.

IV. Artificial intelligence and strategic stability

We consider any parameter whose change may disrupt the established balance of power to be a significant factor in the strategic balance. The nuclear weapon is discussed above as a factor that fundamentally altered the strategic situation. It is known that the kill radius (for devastation) R is proportional to $N^{1/3}$ where N=explosive yield. With the advent of nuclear weapons in the 1940s, explosive power increased by a factor of 10^6–10^7. This made it possible to devastate target areas of hundreds of square kilometres.

But the problem has one more dimension. During the Second World War, the artillery and air force often caused mass destruction by fire. However, target identification and aiming were far from precise. With non-nuclear explosives the circular error probable (CEP) deviation R_0 was significantly larger than the radius of devastation R; that is why the effectiveness of fire was very low. The advent of nuclear weapons reversed the situation: the kill radius R became significantly larger than the CEP, R_0. Taking account of the fact that the probability of hitting the target is proportional to the ratio N/R^3_0, we can see that, alongside increasing the explosive power, there is another way to increase the effectiveness of fire, namely the reduction of the CEP deviation. Of course, this requires much more accurate data on the target's co-ordinates as well as more precise guidance systems. Military applications of AI systems are being developed in just that direction. The cruise missiles equipped with TERCOM guidance systems are already able to hit small targets at distances of thousands of kilometres with a high degree of precision.

Thus a situation is created in which increasing the accuracy of conventional weapons becomes equivalent to increasing their explosive power up to the level of that of nuclear weapons. This trend will be strengthened with the development of AI-based battlefield command and control systems. At present such an approach to combat operations constitutes an important part of the plans of employment of the US Armed Forces known as the 'Rogers plan' which has been adopted in Western Europe as an integral part of the NATO military doctrine.[23] Under these circumstances the AI systems are achieving the status of one of the major factors of the power equation.

Experts often suppose that a possible future world war may begin with non-nuclear strike and only later convert into nuclear-armed conflict.[24] Some experts also believe that the saturation of weapon systems with electronic components may increase stability and create a temptation to begin operations with 'smart weapons', counting on their ability to destroy enemy nuclear arsenals by a preventive non-nuclear strike, The erosion of boundaries between non-nuclear and nuclear weapons with the advent of 'intelligent' systems has a dangerous destabilizing effect on the world military–political situation, especially since it increases the uncertainty in crisis situations.

The process of the diffusion of computing technology into every sphere of society is irreversible. However it can be managed. It is unrealistic to propose banning AI military applications under present political circumstances. Now, it is especially important to concentrate efforts to elaborate confidence-building measures as an integral part of the international security system. A major step in this direction is the appeal of the member countries of the Warsaw Treaty Organization to the member countries of NATO and to all European countries proposing a system of confidence measures to eliminate the fear of surprise attack against any European country, and to reduce the accumulated suspicion and mistrust.[25]

We consider it to be a major responsibility of the scientific community to clarify the possible impact of advanced artificial-intelligence systems on weapon capabilities and on the world military–political situation. It is extremely important to provide objective assessments as opposed to assertions of political and scientific interest groups connected with the military–industrial complex.

Notes and references

[1] *A Complex Programme for Scientific-Technical Progress of CMEA Member States Until the Year 2000* (Politizdat: Moscow, 1985).

[2] Snow, C. P., *Two Cultures* (Progress: Moscow, 1973).

[3] Statement made by Mikhail Gorbachev, General Secretary of CPSU Central Committee, on Soviet television.

[4] Frunze, M. V., *Selected Works*, Volume 2 (Moscow, 1957), p. 367.

[5] Machiavelli, N., *The History of Florence* (Nauka: Moscow, 1973); *Letters about the Use of the Study of History* (Nauka: Moscow, 1978).

[6] Marx, K. and Engels, F., *Collected Works: Volume 20* (Moscow).

[7] Kiryan, M. M. (ed.) *Military-Technical Progress and the USSR's Armed Forces: Analysis of the Development of Armaments, Organization and Methods of Action* (Voenizdat: Moscow, 1982), p. 329.

[8] See note 7, p. 16.

[9] Gareev, M. A. and Frunze, M. V., *A Military Theorist* (Voenizdat: Moscow, 1985), p. 327.

[10] *Krasnaya Zvezda*, 9 July 1986.

[11] Gareev (note 9), p. 237.

[12] Note 7, pp. 281–82.

[13] Winston, P., *Artificial Intelligence* (Addison Wesley: Reading, MA, 1977).

[14] Antonyuk, B. D., *Information System Within Control* (Radio i Svyaz: Moscow, 1986).

[15] Shenk, P., *The Conceptual Processing of Information* (Energiya: Moscow, 1980).

[16] Sergeev, V. M., 'Artificial intelligence as a method of investigation of complicated systems', *System Investigations* (Moscow), 1984.

[17] *A Large-Scale Anti-Missile System and International Security*, Report of The Committee of Soviet Scientists for Peace Against the Nuclear Threat (Moscow, 1986), pp. 48–56.

[18] Venda, V. F. and Lomov, B. F., 'Interaction of man and computer and problems of a cognitive process', *Philosophical Questions in Technical Knowledge* (Nauka: Moscow, 1984), pp. 196–211.

[19] Orfeev, Y. F., 'Concerning formal and informal components within solution of tasks by a man and a computer', *Cybernetics and Modern Scientific Cognition* (Nauka: Moscow, 1976), pp. 333–45.

[20] Pospelov, D. A., *Imagination or Science? Towards Artificial Intelligence* (Nauka: Moscow, 1982), p. 221.

[21] Pospelov, G. S., *Systems Analysis and Artificial Intelligence for Planning and Control: Practical Cybernetics* (Nauka: Moscow, 1984), p. 149.

[22] Wiener, N. A., *Creator and Robot* (Progress: Moscow, 1986).

[23] Semeiko, L. S., 'Back from the brink', *USA—Economics, Politics, Ideology* (Moscow), no.

11 (1986), pp. 75–80; Kokoshin, A. A., 'The Rogers Plan: alternative defence concepts and security in Europe', *USA—Economics, Politics, Ideology*, no. 9 (1985), pp. 3–14.

[24] Gareev (note 9), p. 237.

[25] *Meeting of the Political Consultative Committee of the Warsaw Treaty Organization Member States, Budapest, 10–11 June 1986* (Politizdat: Moscow, 1986).

Part IV. Applications in arms control analysis

Chapter 9. Computer applications in monitoring and verification technologies[1]

TORLEIV ORHAUG

I. Introduction

Space technology was partly developed for military purposes and spaceborne observations rapidly became important tools for surveillance and targeting. Such observations are also used for verifying (bilateral) treaties. The importance of this technology is indicated by the SALT agreement in which measures are taken to prevent interference with 'national technical means' (which are understood to include satellite observations).

Satellite observations of sufficient geometric accuracy for monitoring and verification tasks require ground resolution better than, say, 1–3 metres. Such detailed satellite data are still the monopoly of the superpowers. Several suggestions and studies have been made to use satellite observations for monitoring and control, either in an international or in a regional framework, and many proponents have argued that satellite observations could be a valuable instrument for treaty verification, crisis monitoring and disarmament control.

Applications of satellite observations for these purposes would require the acquisition of imagery data; either from passive sensors (visual, infra-red (IR), thermal IR, microwave radiometry) and/or from active sensors (microwave radar, laser radar). The efficiency of such observations does not only depend upon the timeliness and quality of data; the expediency of data handling, data processing and data analysis is also a factor of great importance in the overall efficiency of a monitoring system.

Problems related to the definition of sensor types, sensor parameters and observational requirements are not examined here. It is assumed that a verification body acquires data relevant for verification purposes with ground resolution of the order of 1 m (or better). The purpose of this paper is to address the problems of image processing/image analysis. The field of computer image processing is reviewed in section II by defining various processing tasks; this is done by considering various 'transformations' of data. It is pointed out that although *image description* is possible using available techniques, *scene description* (image interpretation) is much more difficult to implement in computerized form. The problems of scene description, particularly 3-D (dimensional) scenes, are analysed in some detail. The state of the art in scene analysis (computer vision) is briefly described and the problems of introducing context and other kinds of knowledge in computer vision are pointed out.

Recent research in the field of artificial intelligence (AI) is of particular importance for computer vision. Section III discusses how AI methods can be applied to image analysis. The state of the art is briefly described and current research areas are indicated.

Next, the problem of processing speed is addressed in section IV. It is shown that for institutionalized verification programmes, a vast amount of imagery data must be processed; high speed processing is, therefore, mandatory. Various ways of increasing speed are discussed, such as algorithmic smartness, and improved computer architecture and hardware implementation. Also, the division of processing load between on-board processing and ground processing is indicated.

Section V treats the various processing tasks in verification using satellite imagery and indicates the role of and the need for computer processing. For this discussion, a 'model' verification centre is assumed with a defined number of images to be processed and interpreted. The various processing tasks which can be carried out routinely using available computer technology and available processing methodology are identified and those areas in which intensive research is still going on are also indicated. Although computers will aid in interpretation, the need for a staff of highly trained photo-interpreters for the final analysis and judgement is of paramount importance.

II. Computer image processing

The development of technical means for image acquisition and image generation dates back some 100 years to the invention of photography. Photographic techniques have been used for image manipulation/processing such as enhancement, pseudo-colour coding, contrast variation, and so on. The applications of these methods are, however, very limited. The same has been true for other analog techniques, television technology being the first electronic technique used for image processing. During the 1960s, new techniques in terms of coherent optics/holography and digital computers became available and the latter still comprise the fundamental technology base for image processing.

In the theoretical domain, many of the processing schemes were based upon the well-established area of signal theory (dealing with time signals). The limitations of such techniques later became evident and image analysis now also benefits from other fields such as visual perception and artificial intelligence, in particular that branch of AI dealing with computer vision/signal understanding. Developments and investments in the field have been partly technology-driven (developments in computers and computer science) and partly application-driven (automatic photo-interpretation, industrial and military robotics, missile guidance and medicine, to name just a few).

The role of (civilian) remote sensing for the investments in and progress of image processing should also be emphasized. The first large digital data bank to become generally available comprised remotely-sensed satellite imagery data.

This created a significant interest in and need for computer techniques for filing, handling, processing and analysing such data.

Data transformations

A structural framework for the various processing tasks may provide a valuable background for the discussion below. The terms 'image' and 'data' are used as follows: 'image' refers to imagery data in such a form that when displayed in correct order (e.g. line- and column-wise) they give a pictorial representation of a scene with perceptual meaning with regard to geometric characteristics; 'data' on the other hand, although they represent imagery data, are coded such that their display does not have a visual perceptual meaning in the above sense. Although the data represent the scene, they must be decoded or transformed to give a perceptual meaning. Examples of such data are raw SAR-data (SAR: synthetic aperture radar), 2D-Fourier transforms of an image and tomographic data. With these vague definitions in mind the following processing tasks of imagery data or ways of transforming such data can be defined:

1. Image—image: this transformation comprises a variety of processing schemes such as enhancement methods (filtering, crispening, colour coding (pseudo- and false colouring), geometric transformations, contrast manipulations, noise removal methods, etc.).

2. Data—image: this covers transformations of coded imagery data to images (pictorial form); examples are SAR-processing for image generation, image generation in tomography, decryption, and decoding of compressed images.

3. Image—data: this transformation includes linear transforms (2D-Fourier, Hadamard, Karhunen-Loeve, Haar, etc.), encryption and coding schemes such as data compression methods.

4. Image—image description: this transformation covers a large number of methods in low-level processing, such as feature extraction (lines, curves, edges, texture characteristics), but descriptions from coded images (Fourier-transform features) and intermediate-level processing (such as region detection and thematic mapping) also belong to this group.

5. Image—scene description: this is also called image understanding, scene analysis, computer (machine) vision; the scene is often a 3-D scene; description of 2-D scenes is in principle less difficult (compare many problems in medical image analysis and analysis of low-resolution remote-sensing images).

6. Data—image/scene description: extracting information from data; properties of ocean waves directly extracted from raw SAR data belong to this group.

Scene description

The purpose of scene description/computer vision is to extract information (intelligence) from images. Image processing/scene description/image understanding can be considered as part of a broader class: signal processing/signal understanding pattern recognition. Imagery data may be considered to constitute a special sub-class of signals with special properties.

Pattern recognition may be divided into two classes: conventional methods (statistical and structural methods) and AI methods respectively. Simple pattern-recognition tasks may be described as classification (template matching), while more elaborate methods belong to signal understanding. Signal understanding covers procedures where interpretation of the signal is carried out such that the exact structure of the recorded signal is interpreted ('understood') in terms of the signal-generating mechanism (illumination, aspect and viewing angles and other sensor characteristics for imaging).

It should be emphasized that scene description in general always requires more information than is inherent in the imagery data themselves.

Statistical pattern recognition classifies the signal into one of a number of classes based upon statistical techniques (e.g., multivariate probability distributions). The classification scheme utilizes a number of features (which can be compared with multispectral features in remote sensing). This is often called a supervised method (in which class probabilities are known *a priori*). In unsupervised classification ('clustering'), class affiliation (classification) is based upon the clustering tendency of the data. This methodology is application-independent; the same schemes may be used for various types of signal and for various applications. Although the basic technique itself is based upon optimization schemes (optimization of error probabilities or costs), the optimization is arbitrary since the features (the data input) are only specified on an *ad hoc* basis. In remote sensing (thematic mapping) this technique has proven quite successful. It is not as useful for 3-D scenes, however, since statistical modelling of such scenes is not often very meaningful. This is particularly true for monocular images depicting 2-D projections of 3-D objects.

The fundamental problem of 3-D image analysis/scene description is an inversion problem. The image itself is fully determined by the scene and the sensor characteristics. The inversion problem is, however, a highly underdetermined problem; an infinite number of scenes may give the same image. Another way of stating this is that the relationship between the image descriptors (features) and significant scene characters is not known *a priori*. Two examples illustrate this point. Edges may be related to objects or to shadows. The information carried by the existence and position of edges depends critically upon the kind of edges concerned. Textures may be analysed as an image feature (and many techniques are available). Unless a theoretical basis for describing image texture as related to scene properties exists, image texture characteristics cannot easily be used for inferring scene properties. In conventional remote sensing, the relationship between image features (multispectral features) is established by field measurements of spectral characteristics and/or by identifying homogeneous areas with known ground characteristics ('ground truth establishment').

Most of the concepts used in image understanding to date have been based upon *data driven* methods. Features are extracted from the image and, on the basis of these features, homogeneous regions are identified (as in the 2 1/2D sketch according to Marr[2]). Image properties (shading, occlusion, etc.) are then used to infer surface orientations. The next step is to infer and recognize objects and their interrelationships. There is a great need for complementing this approach by hypothesis-driven concepts in which, at various levels in the processing flow, data/information from lower levels are used for deciding between competing hypotheses. Also, in the scheme indicated above, inference of depth in the scene comes late in the processing chain. There are interesting indications from experiments in visual perception that inference of depth comes very early in human vision. If and how this should influence computer vision is one of the many problems now being investigated.

III. Scene description and AI

The AI concept cannot readily be defined; the methodology uses descriptions of abstract concepts represented by several levels of abstraction (compare low-, intermediate- and high-level processing in vision) and the recognition of instances of such concepts in the signal. At each level of abstraction in the hierarchy, knowledge appropriate to that particular level is used to identify components of the higher level of abstraction and vice versa. Also, at the different levels of abstraction, different knowledge sources are used to model the mechanism that generates and deforms patterns. The AI approach is concerned with the representation of concepts and with procedures to identify their instances in the signal domain. An important feature of AI systems is that they use *explicitly* coded heuristic information to generate hypotheses about the existence of a concept and to evaluate the hypothesis.[3] Knowledge is normally represented by semantic networks and schemes. In the former method information is represented by a set of connected nodes; the connections are made by labelled arcs (representing relationships among the nodes). Algorithms are required to generate and evaluate inference. Such methods have been used for interpreting aerial maps. Inference is hypothesized and evaluated by the knowledge source at various levels and the structure of the knowledge sources may be different at different levels. Finally, a control mechanism is used to decide the strategy for understanding and also for efficient utilization of the knowledge source; this also provides user interaction.

An AI *expert system*, as described above, uses the heuristic knowledge of a domain expert in combination with decision rules to achieve the desired goal. The problem of 'knowledge engineering' is to have the expert explicitly describe the knowledge he uses in his interpretation; it is normally not possible beforehand to specify all the heuristics needed for the analysis.

In an expert system, the knowledge source is represented as a series of *production rules* and these can be represented as a series of logical statements of the form: 'if . . . then'. For example if (P1 and P2 or P3) then (P4). The predicates (the Ps) can be either facts or conclusions from previously applied production rules; P4 in the example above can be used as a predicate in some other procedure at a higher level of abstraction. The production system has three concepts: the rule base, the data and the control structure; the rule base should be open-ended thus facilitating the addition or subtraction of information. The data used in the system consist of the predicates (primitives—low-level input information) and derivatives from production rules. The applications of the production rules are accomplished by the control structure.

The important characteristics of an expert system may be summarized as follows:

(*a*) the system separates domain knowledge from the way this knowledge is used;

(*b*) the system incorporates knowledge in a flexible way and knowledge may easily be added; and

(*c*) the system can be made transparent thus describing the manner it has used to arrive at a certain conclusion and the data supporting this conclusion.

These characteristics are in contrast to 'conventional' image analysis systems in which knowledge is embedded (implicit) in the algorithm (compare multispectral classification or low-level labelling for edge extraction). It should be pointed out that an expert system for image understanding/computer vision will contain many of the image-processing tools used in conventional image analysis.

During recent years, expert systems have been applied to imagery data on a research basis. Examples are the ACRONYM system for mapping image properties such as edges to instances of object models. A system for segmenting and interpreting colour-IR aerial photographs showing roads, rivers, forests, residential and agricultural areas has also been described.[4] Another system has been used for guiding interpretation of suburban residential scenes in monochromatic aerial photography.[5]

Because of the difficulties of implementing automatic interpretation of single imageries another branch of research is directed towards using data from different sources (imageries) as data input to an expert system (*data fusion*). The idea is that by using combinations of information describing very different aspects of objects, the interpretation for each single image should not need to be carried out at a very high level of abstraction. Examples of imagery data used as input are optical images, thermal IR, radar with moving target indicator detection (establishing the location and velocity of objects) and microwave signal detection (communication and/or radar signals). Data fusion is considered to be very important in tactical military situations where speed is important.

It should be pointed out that current expert systems contain a frame for utilizing data and information in a flexible manner. The systems are not, however, explicitly designed or easily adapted for utilizing and interpreting imagery data from 3-D scenes. Spatial data and spatial knowledge have many aspects which are difficult to verbalize and an efficient language for expressing such data does not exist as yet. Examples of such aspects are constraints on shapes and spatial relations. In particular, expert systems offer few techniques of relevance for low-level vision. The problem of inferring properties of scene geometry are not explicitly solved by present systems.

IV. Image processing and speed

The technological ability to acquire imagery data is far in excess of the ability to process and analyse such data (automatically) by technical means. The present Landsat system acquires an image consisting of approximately 200 Mbytes of data every 30 seconds. If only the global land masses are sensed during daylight, the imagery data output is annually approximately 8 Tbytes of data consisting of roughly 400 000 single images.

The huge data flow from observation–satellite systems means a significant data-processing load for the ground stations. Three, somewhat different methods, may be used to handle this load:

1. Distributing the processing load between on-board and ground processing;
2. Increasing the (ground station) processing speed by: (*a*) improving the speed of processing elements; (*b*) designing efficient algorithms; and (*c*) developing better architecture.
3. Screening imagery data in order to process only 'relevant' data.

These approaches are briefly commented on below.

On-board versus ground processing

On-board processing can only be used for 'simple' image-processing tasks such as correction, data compression, registration, and so on, and for simple analyses such as hot spot detection (compare superpower monitoring of launches of intercontinental ballistic missiles (ICBMs) with strategic monitoring systems), multispectral analysis, and so on. Nevertheless, such processing enables the screening of interesting regions which should be subject to detailed investigation and are therefore important to a monitoring agency.

The compression technique is the one which has been most thoroughly examined and a recent study shows a compression ratio of approximately 2:1 to be feasible for the Advanced Landsat Sensor (lossless compression).[6] Data compression is particularly needed to decrease the channel requirements for data transmission.

Data analysis, on the other hand, is a more open question since on-board processing is application-dependent.

Processing elements

The main processing elements available are the optical (analog or digital) and solid-state elements used in digital (electronic) computers. The optical computer is not yet flexible enough to offer real competition although its operating speed (in principle determined by the speed of light) is superior to that of the digital computer.

Solid-state technology may be based upon the standard silicon-based integrated circuits, gallium arsenide (GaAs) and Josephson devices; the first two are normally operated at room temperature. The speed of present silicon circuits is of the order of 10–100 Mops (mega-operations per second). Because of the more complex techniques involved in GaAs-circuitry the corresponding chip design lies some eight years behind the well-proven silicon devices. Today, great interest is focused upon low-temperature techniques. Some computers are already based upon liquid nitrogen cooling (77°K), increasing speed by approximately seven times. If liquid helium (4°K) or liquid nitrogen (77°K) is used, superconductive effects may be exploited. Interesting candidates are Josephson devices which may be 1000 times faster (e.g., for analog/digital conversion) than ordinary room-temperature devices.

Efficient algorithms

Design of efficient and clever algorithms often implies taking short cuts in a program and unfortunately leads to non-transparent programming.[7] In special cases, efficient programs have nevertheless increased computation efficiency by orders of magnitude; examples are Fast Fourier Transform (FFT) and matrix transposition.[8]

Computer architecture

The basic method of speeding up computation is parallelism. Using n processors a degree of hardware parallelism is implemented and provided these processors operate concurrently, speed is enhanced by a factor of n. In traditional sequential (von Neumann) computers, n=1. The various types of hardware parallelism have been categorized by Danielsson and Levialdi[9] and the types include: (*a*) pixel-bit parallelism, (*b*) neighbourhood parallelism, (*c*) operator parallelism, and (*d*) image parallelism.

In pixel-bit parallelism, each pixel is coded as a computer word and enhanced speed is obtained by word processing instead of bit processing. In neighbourhood parallelism, the processor computes one output pixel depending upon the neighbourhood of the input pixel. Operator parallelism is equivalent to pipelining ('focal-plane architecture'). In this method each data set is passed from one set of processors to another. In image parallelism, one processor is connected to each pixel in the image. Processing can thus be carried out at the same speed over the whole image simultaneously. A number of different computer architectures have been suggested. The following are (representative) examples:[10]

(a) the sequential machine (von Neumann architecture);
(b) the single instruction multiple data stream (SIMD-architecture or array processors);
(c) pipeline architecture; and
(d) the paracomputer.

The first type of architecture represents the 'classical' computer still in use for general purpose computing. The array processor uses a set of processors executing the same instruction and is driven by a host computer. This processor is well adapted for vector processing. The pipeline machine is well suited for neighbourhood processing. The paracomputer (utilizing full image parallelism as described above) was introduced by Schwartz[11] and is a concept well suited to performance evaluation since it represents an 'ideal machine' for a given speed of the processor elements. The number of processors is equal to the number of pixels in the image.

In addition to the above architectures, a number of others are being studied such as MIMD (multiple instruction multiple data stream), Multi-SIMD and so on. Special-purpose hardware functions are also becoming available (e.g., systolic arrays). These architectures have been implemented for special tasks such as FFT, low-level processing and so on. Often, such hardware equipment is embedded in large systems or used for real-time applications.

Most of the commercial image-processing systems marketed today are of the SIMD-class. An illustration of the increase of computing power since 1960 has been given by Preston, evaluating several commercially available computers designed for biomedical image processing.[12] Figure 9.1 gives the number of picture point operations per second (pixops) for several systems. The pixop rate increased from 10^3 in 1960 to 10^{10} in 1980; an increase of one order of magnitude every three years.

An interesting comparison of various machine architectures has been published by Cantoni et al., who computed execution time and communication time (data-loading time plus instruction-loading time) for several types of operation for the four types of architecture listed above (sequential, SIMD, pipeline and paracomputer architecture).[13] The following operations were investigated: point operations, local (neighbourhood) operations, image statistics (histogram), co-occurrence matrices and 2-D Fourier transforms. The image size investigated was 128×128 pixels and the neighbourhood size was 3×3 pixels. The clock cycle and the execution time were 300 nanoseconds and 1 microsecond respectively.

The results of the (theoretical) investigation are shown in figure 9.2 giving the computation and communication times for each architecture type and each

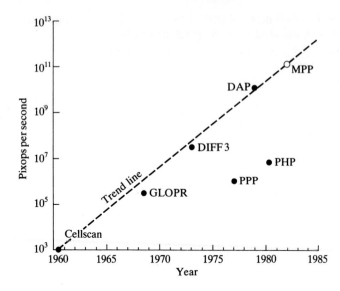

Figure 9.1. Picture point operation (pixop) rates for various parallel-image processing computers
Source: Redrawn from Preston, K., Jr, 'Comparison of parallel processing machines: a proposal', in Duff, M. J. B. and Levialdi, S. (eds), *Languages and Architectures for Image Processing* (Academic Press: London, 1981), p. 307.

operation respectively. These results are used below in examining computation time for verification studies. An interesting observation is that only the SIMD-architecture is 'communication limited'. For the other two schemes, execution time is greater than communication time.

V. Image analysis for verification

An image-interpretation facility using digital imagery from satellite-borne sensors would use computer processing for the following tasks:

1. Image calibration and standardization—this covers both radiometric and geometric transformations accounting for sensor and platform effects to produce 'standardized' images; also, effects of variations in illumination (resulting from solar angles caused by diurnal and seasonal effects and topography) should be compensated for; this is important both for image comparison (change detection) and for object detection.

2. Change detection—the detection of changes either between image and map or between images; precise geometric and radiometric standardization is needed—this is not easily achieved automatically; this detection could also be carried out at the 'object' level provided the difficult task of object detection has already been performed.

3. Object detection—automatic/interactive detection of a number of poten-

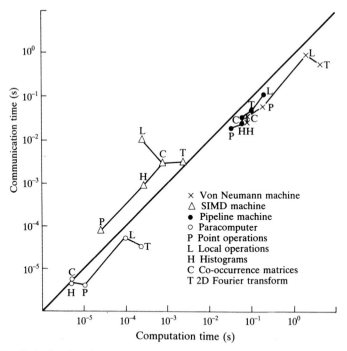

Figure 9.2. Relationship between computation time and communication time for several computer architectures for some given image-processing tasks
Source: Redrawn from Cantoni, V., Guerra, C. and Levialdi, S., 'Towards an evaluation of an image processing system, in Duff, M. J. B. (ed.), *Computing Structures for Image Processing* (Academic Press: London, 1983), p. 55.

tial targets such as urban–industrial infrastructure (roads, ports, airfields, industry, etc.), strategic/tactical targets (missile sites, vehicles, ships, etc.).

The image-processing algorithms needed for the above tasks cannot easily be listed. A whole repertoire of algorithms is needed for (*a*) line- and edge-detection/tracking; (*b*) multispectral classification; (*c*) linear transforms; (*d*) geometric transforms; (*e*) convolution and filtering; (*f*) texture algorithms; (*g*) relaxation; (*h*) pyramid structure generation; (*i*) feature extraction; (*j*) model matching and (*k*) map matching. For object detection, for example, several algorithms must be used, such as those for line and edge detection/tracking, area detection using pyramids, matching, and so on.[14] Also, combinations of several techniques are needed for interpretation, such as (*a*) visual image inspection of hard copies and displayed images; (*b*) interactive computer-assisted techniques; and (*c*) computer-image interpretation using knowledge-based systems.

A model image analysis facility

An image analysis facility should contain: (*a*) facilities for mass storage for archiving both imagery and (digital) map data; (*b*) large imagery data bases for fast access; (*c*) large data bases (for non-imagery data); (*d*) general-purpose image-display and image-analysis computers with many work stations incorporating large image display memories for fast screening through imagery data with quality displays; (*e*) computer systems facilities for geographical information systems (GIS); and (*f*) computer systems incorporating GIS and vision expert-system facilities.

As an illustrative example a 'model image analysis facility' is considered charged with the following responsibilities:

1. The control/monitoring of a geographical region of 10^5 km^2 (of the order of magnitude of a central European monitoring scenario).
2. Reception of imagery data from monitoring satellites at an average rate of 100 images per day, each image of the size of 10^8 pixels in 4 spectral bands giving 40 Gbytes per day. One pixel is assumed to cover a square metre on the ground.
3. Pre-processing of the imagery data at the ground station (the technical and infrastructural facilities needed for this are not considered here).
4. Routine inspection of the acquired images for signs of changes of importance to further detailed inspection for precise detection of changes and control of activities.

Two different methods are available for estimating computational load: detailed description of the necessary algorithms for evaluating execution time, and estimation of 'average' or typical algorithm complexity for such evaluation. The latter method is used here. Also, since the execution time for point operations is at least one order of magnitude shorter than local (and global) operations, the influence of point operations is neglected. In order to estimate execution time it is assumed that:

1. For object detection (or similar computer-costly operations) the execution time needed is equal to that for 20 local operations
2. The local operations (which will be executed on images having different resolution—using resolution pyramid representation) should have a maximum size of 50×50 pixels
3. The execution (and communication) times needed for various types of architecture can be scaled linearly (according to image and operation size) from the data given by Cantoni et al.[15]

Further it is assumed that for the various computer architectures:

1. The number of processors are: von Neumann, n=1; SIMD, n=10^4; pipeline, n=10^4; paracomputer, n=10^8.
2. The number of machine instructions (to execute the algorithm) is 100.

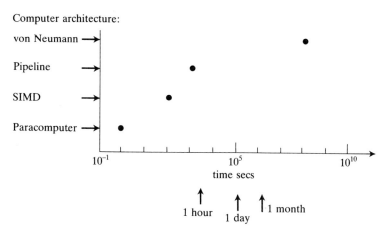

Figure 9.3. Approximate total execution time for processing 100 images for verification analysis
Source: Based on data published in Cantoni, V., Guerra, C. and Levialdi, S., 'Towards an evaluation of an image processing system', in Duff, M. J. B. (ed.), *Computing Structures for Image Processing* (Academic Press: London, 1983).

The corresponding total execution time for scanning 100 images (neglecting communication time) is given in figure 9.3. The total execution time lies between 1 second for the paracomputer and 10^8 seconds (approximately three years) for the sequential computer. The interesting schemes of SIMD and pipeline computers both have execution times of the order of one hour.

These time estimates should be treated with extreme caution since they do not give a realistic estimate of the computer time needed for image analysis for verification purposes. Image-loading, access, communication, interaction, and so on, are not taken into account in this analysis.

Nevertheless, the example indicates what developments in computer architecture can achieve, making it feasible to interconnect thousands of processors in an efficient image-analysis computer structure. It should be emphasized that even if 'automatic' computerized image analysis can give indications about single or multiple objects, analysis and interpretation at higher levels of abstraction—such as type of military unit, organization and intention—are very difficult tasks requiring contextual and background information. For such tasks the skill of photo interpreters/domain experts must work in an interactive mode with various kinds of image analysis computer.

VI. Conclusion

During the past 20 years computerized methods for image processing and image analysis have been developed and computer methods have become important tools for various tasks in dealing with imagery. Nevertheless, image interpretation adequate for arms control verification purposes cannot yet be carried out automatically without the detailed interaction of domain experts. On the other hand, computer methods will become more and more important for verification tasks and various computers, including knowledge-based systems, will be necessary for efficient image interpretation.

Notes and references

[1] The author acknowledges helpful discussions with S. Nyberg of the Swedish National Defence Research Institute and with Dr B. Kruse at Teragon.

[2] Marr, D., 'Representing visual information in computer vision systems', in Hanson, A. and Riseman, E. (eds), *Computer Vision Systems* (Academic Press: New York, 1978), pp. 61–80.

[3] Nandaklemar, N. and Aggarwal, J. K., 'The artificial intelligence approach to pattern recognition—a perspective and an overview', *Pattern Recognition*, vol. 18, no. 6 (1985), pp. 383–89.

[4] Bullock, B. L., *et al.*, 'Image understanding application project: status report', *Proceedings of the DARPA Image Understanding Workshop*, Palo Alto, CA, 15–16 Sep. 1982, pp. 29–41.

[5] Perkins, W. A., Laffey, T. J. and Nguyen, T. A., 'Rule-based interpreting of aerial photographs using LES', *Proceedings of SPIE*, vol. 548 (9–11 Apr. 1985), *Applications of Artificial Intelligence*, pp. 138–146.

[6] Schueler, C., de Boer, C., Marks, B. and Stegall, M., 'On-board data compression for advanced Landsat', *Proceedings of SPIE*, vol. 534, *Architectures and Algorithms for Digital Image Processing II* (1985), pp. 135–141.

[7] Davies, E. R., 'Image processing—its milieu, its nature, and constraints on the design of special architectures for its implementation', in Duff, M. J. B. (ed.), *Computing Structures for Image Processing* (Academic Press: London, 1983).

[8] Eklundh, J-O., 'A fast method for matrix transposing', *IEEE Transactions on Computers*, vol. C-21 (1972), pp. 801–803.

[9] Danielsson, P-E. and Levialdi, S., 'Computer architectures for pictorial information system', *IEEE Transactions on Computers*, vol. 14, no. 11 (Nov. 1981), pp. 53–67.

[10] Cantoni, V., Guerra, C. and Levialdi, S., 'Towards an evaluation of an image processing system', in Duff, M. J. B. (ed.), *Computing Structures for Image Processing* (Academic Press: London, 1983).

[11] Schwartz, J. T., 'Ultracomputers', *ACM Transactions on Programming Languages and Systems*, no. 4 (1980).

[12] Preston, K., 'Comparison of parallel processing machines: a proposal', in Duff, M. J. and Levialdi, S. (eds), *Language and Architecture for Image Processing* (Academic Press: London, 1981).

[13] See note 10.

[14] Bogdanowicz, J. F., 'An evolving system for image understanding', *Proceedings of SPIE*, vol. 528, *Digital Image Processing* (1985), pp. 110–16.

[15] See note 10.

Chapter 10. Knowledge-based simulation for studying issues of nuclear strategy

PAUL K. DAVIS

I. Introduction

This paper is concerned with methodologies for studying three issues in nuclear strategy: deterrence, escalation control and war termination (DEWT).[1] A premise here is that current methods are not only inadequate, but in some instances pernicious in that they focus attention on the wrong issues, pose the wrong choices and lead to foolish or even dangerous policies. The problems have not gone unnoticed all these years, for some of the best minds available have worked on nuclear strategy. However, we are all affected by the methodologies of our era, and the tendency is to use them even if they are manifestly unsatisfactory until such time as there is a good and convenient alternative. The author believes that better methods are now becoming available. They will be no panacea, but they will help fill important gaps that have caused trouble in the past.

What, then, are the methodologies for which improvements may be emerging? On the quantitative side of allegedly hard analysis, they include: (*a*) cataloguing 'static measures' (e.g., numbers of launchers, warheads, throw-weight and so on), (*b*) performing 'exchange calculations' (e.g., calculations describing results of hypothetical missile duels or, in more sophisticated simulations, the results of the United States and the Soviet Union launching their full nuclear arsenals), and (*c*) performing system-by-system comparisons in some technical depth (e.g. comparisons of missile accuracy, reliability, and range).

On the non-quantitative side it is difficult to identify 'methodologies' *per se*, but there is a rich strategic literature on DEWT issues that includes classic efforts by Brodie,[2] Schelling,[3] Kahn,[4] Wohlstetter,[5] and Ikle.[6] It is to this literature that we can trace such basic elements of our thinking about DEWT issues as: (*a*) *escalation ladders*; (*b*) escalation control as a form of *bargaining* with explicit or tacit thresholds; (*c*) the role of mathematical game theory in understanding dilemmas in crisis; (*d*) the role of *scenarios* in exploring alternative futures; and (*e*) various mathematical and non-mathematical models of crisis stability.

At first glance, this list would suggest that we have a great deal to work with in our studies. And, indeed, that is true. However, despite the brilliance of some of this early work there have also been some serious shortcomings.

1. A failure adequately to treat asymmetries of mind-set between West and East, and indeed among NATO alliance members;

2. A failure to recognize adequately the extraordinary confusion faced by real-world decision-makers working within a maze of organizations to cope with problems that are only poorly understood—i.e., a failure to deal significantly with bounded *rationality*;

3. An almost exclusive focus on *offence-dominated strategic nuclear force postures*;

4. Inadequate attention to operational and political–military constraints; and

5. The tendency to encourage *trivialization* in search of mathematically tractable or otherwise simplified problems, and the tendency to leave 'rigorous' analysis to quantitatively oriented workers with inadequate sensitivity to issues of political science, psychology, and decision-making theory.

None of this should be construed as criticism of the original work, which continues to be important today. However, we are now more than half-way through the 1980s and more should be demanded of our methods and theories than was expected in the 1960s.

All of the shortcomings mentioned above have been recognized and discussed over the past 15 years. However, in the view of the author, much of the discussion has come to much less than might have been expected. Instead, it seems that we find ourselves in a world of competing but non-convergent essays. That is, there is no dearth of essays or books discussing strategic theory generally, asymmetries, organization theory, and so on, but neither is there any analytical framework for comparing views rigorously and moving towards conclusions—especially when the comparisons cross disciplinary boundaries.

A basic problem here is the inadequacy of prose for comparative analysis of complex issues—the nature of the medium encourages authors to work with stereotypes or to lapse into heavily caveated and balanced discussions ('On the one hand . . . but on the other hand . . .'), which are often stimulating but which seldom lead to sharp conclusions. The standard prescription for improving rigour in complex problems is to develop analytical models—not vague models or taxonomies, but operational models with inputs and outputs, models that can help analysts reach conclusions for a particular context. A discussion of how the Rand Strategy Assessment Center is moving towards building such models is given below.[7]

II. Breaking out of a cultural heritage

There are some interesting cultural factors that have delayed development of some of the ideas most necessary for the work discussed below. These have been well noted and discussed by Prigogine in the book *Order Out of Chaos*.[8] A major point of Prigogine's argument is that, by and large, our western culture strongly values physical or quasi-physical laws such as those of Newton, which describe relatively simple idealized systems. Problem-solving tends to be done by starting with a simple idea, developing rigorous solutions, and then extrapolating—sometimes outrageously—to real systems. This approach has served us well over the centuries, and has surely been a prime factor in the successful development of modern physics, chemistry and engineering. It also underlies much of modern economic theory. And, in the realm of national security analysis, it underlies the paradigm of systems analysis as it was defined in the 1960s. Some representative examples of such thinking include: Newton's Laws, Maxwell's Laws, the ideal gas model, the theory of the Bohr atom, the profit-maximizing law of some economic theory and the missile-dual calculation methods of strategic analysis. In all of these instances, the model is in some sense simple, understandable, quantitative and precise. Also, the model deals with *isolated* systems.

Let us consider a well-known model for good decision-making, popular in the West: the use of decision trees and the maximization of a utility function.[9] This technique, which is often called decision analysis as though there were no other way to analyse decision options, is again simple, elegant, useful, and insightful. Again, the approach tends to focus on reductionism, quantification and idealization.

Suppose, now, that one is tasked with analysing deterrence, escalation control or war termination in a society in which such models and techniques are common and revered. Would it be surprising to see analysis with some of the following features?

1. Equating nuclear deterrence with having a second-strike capability to destroy some fraction of the opponent's industrial capacity;

2. Equating crisis instability with mathematical constructions such as the ratio of nuclear weapons that would obtain after a first strike (if, by going first, one side can materially improve the ratio of nuclear weapons it is often said that the posture is crisis unstable);

3. Equating the strategic balance with the net effect of comparing one or a number of static indices such as the number of intercontinental ballistic missiles (ICBMs), the number of submarine-launched strategic missiles (SLBMs), and so on;

4. Analysing potential national dilemmas in crisis by means of idealized game-theory problems such as the famous Prisoner's Dilemma;

5. Analysing the value of options such as Launch Under Attack in terms of the difference in post-exchange weapon ratios depending on whether one does or does not launch under attack;

6. Whenever possible, imputing to each side of a two-sided computer battle the 'optimal behavior' of a game-theoretic solution based on perfect information and something like a minimax strategy.

The intention here is not to criticize these features of our analysis so much as to comment on them. Indeed, the present author has written many papers using all of the above for strategic analysis. For some purposes these methods are both appropriate and effective in laying bare fundamental realities about which we should all be aware.

Unfortunately, the methods—and our tendency to love them—have important drawbacks. In particular, the reality is that not all problems are simple and not all systems are 'close' to ideal systems that can be modelled neatly. Nor are all systems 'static' or 'isolated'. By thinking too exclusively about the idealized problems, there is a tendency to overlook issues of first-order significance in real systems.

A good example here is command and control—command and control in the most general meaning of the words. In probably 90 per cent of all discussions about command and control in the United States, the subject is really the communication *systems*—i.e., the technological parts of the system that can be seen, counted, evaluated and represented in quantitative models. However, to a military historian, or to a general officer who has not been co-opted by technological pursuits, the command and control system consists primarily of organizational structures, procedures and doctrine, with the communication systems serving merely to provide critical information and disseminate the conclusions. The essence of command and control is not in the entities that lend themselves so well to quantitative calculations, but in the 'softer' aspects of the system—the parts for which one needs to understand the social sciences.[10]

In summary, then, despite the background of excellent and insightful work on strategic issues, there remain many serious problems. To a significant extent, these problems reflect our cultural heritage, which has placed a premium on problem-solving based on simple models of simple systems. The relatively new developments that make it possible to address the problems effectively and, happily, to do so in a way that is comfortable within our culture are discussed below: while the new methodologies are not 'simple', they can appeal to us through their use of high technology, which Westerners, and especially Americans, enjoy greatly.

III. Relatively new developments permitting new methodologies

Simulation

One consequence of the computer era has been the emergence of simulation modelling as a technique for studying problems ranging from the understanding of molecular phenomena in dense gases to the prediction of weather and stock-market performance. Intellectually, simulation is very different in nature from the more classic techniques of problem-solving involving closed-form equations. To illustrate, consider the problem of describing the height of an object thrown upwards. From Newton's Laws we find that:

$$H(t) = H_0 + V_0 t - (1/2) g\, t^2$$

where the terms on the right-hand side represent the initial height, the contribution of the initial speed provided by throwing the ball, and the effect of gravity, which eventually brings the ball back to earth. Note that what we sought, H(t), is all alone on the left side of the equation and that finding H(t) merely requires evaluating the right side. This 'closed-form solution' is possible by virtue of simple integral calculus and a number of major assumptions such as: (*a*) there is only one ball throughout the time of interest (i.e., the ball does not, on occasion, spew out daughter balls); (*b*) the ball does not change character in mid-flight; (*c*) the regime in which it flies does not change discontinuously with a change in the gravitational constant; (*d*) the force of gravity is not affected by the ball itself, and so on. Other assumptions can be made, for example by ignoring the friction of air, but those are irrelevant to our discussion since they still permit a closed-form solution, albeit one with an integral sign on the right-hand side.

Consider next the problem of predicting the locations over time of many idealized balls bouncing around within a box—hitting and rebounding not only from the walls but also from each other, with no loss of energy. Newton's Laws of physics still apply, but trying to solve the corresponding equations to produce something simple like $X(t)=f(t)$, where $X(t)$ is the X position of the first ball and $f(t)$ is some straightforward function of t, does not work, because while the position of the walls may be constant, everything else (except total energy and momentum) is constantly changing as collisions occur: where ball X will be shortly depends not only on where it is now, how fast it is going, and constant factors like the position of walls and gravity, but also on non-constant factors such as whether there will be a collision with another ball.

While developing closed-form solutions for such many-body problems is typically impossible, it *is* possible to predict the time history of events using simulation techniques. Here one breaks the problem up into intervals of time (the intervals between collisions). Within each of these intervals, the movement of the balls is easily predicted with closed-form expressions. The results of the collisions can also be predicted (for idealized hard balls) with conservation laws, and one then has another collisionless interval, and so on. In a computer, then, one can replicate the entire sequence of events and produce predictions of where all the particles will be over time. As the number of particles grow, the calculations become more demanding, but that is 'merely' a problem of computer performance.

Let us next consider an example of simulation in the social sciences. A continuing issue of human societies is how best to assure that citizens have homes in which to live. There continue to be debates on the subject—with some arguing, for example, that the central government (either local, state, or national) should build the housing and then allocate it, and with others arguing that the free-market system should be used instead, with government intervention to make it attractive for that system to produce the desired results. The second group may argue, at least in some instances, that the well-intentioned effort of a local government to build housing for the poor will make things worse rather than better by attracting more indigent and unproductive people to the community and discouraging industry from opening or expanding factories in the area. A better approach, that group would argue, would be for the government to support development of 'infrastructure' (e.g., roads and railway networks) and to create tax incentives to attract industry, which in turn would provide jobs, which in turn would provide people with money, which in turn would attract builders to create the desired housing and, perhaps, a stronger local tax base for the community to use in subsidizing those who truly must be subsidized.

The present purpose is not to argue for or against any such policy, but rather to note that the character of the second argument depends on a mental simulation of events, and that the mental simulation can be replaced by a computer simulation in which the assumptions can be made rigorous, can be reviewed and subjected to analysis, and so on. There will remain uncertainties, but at least the process of thinking through the consequences of alternative policies can be made subject to rational analysis.

This idea is not really very old. Indeed, they were highly controversial when introduced by Jay Forrester of MIT in the late 1950s and 1960s.[11] In his approach, which he called System Dynamics, Forrester emphasized that to carry out such policy-relevant simulations various and sundry 'soft' issues must be addressed, and explicit assumptions must be made about them, whether there was a good empirical basis for the assumptions or not. He also argued that the conclusions of the simulations were often valuable even in the midst of uncertainty because they would reveal phenomena that, without such analytical aids, people are cognitively ill-suited to recognize: feedback phenomena such as the negative effects of the state building housing in an environment that would then allow other people needing housing to flow into the area. Let it suffice to observe that the application of simulation to policy problems is still relatively new, and that most of the methodology used in discussion of strategic policies was developed before simulation was widely used, and before attention had been paid to Forrester's work.[12]

Gaming

Political–military gaming emerged in the 1950s in the United States, and military gaming has, of course, been conducted for centuries.[13] These activities represented a type of simulation possible long before computers were powerful enough to be a factor. They also had many of the generic advantages of simulations. For example, in political–military gaming one can explore the potential impact of such 'soft' issues as:

1. The reaction time of allies in periods of ambiguous strategic warning;
2. The co-operation of allies and third countries in providing combat forces, airfields, intelligence and logistics; and
3. The many real-world problems that exist in command and control (e.g., poor communications, misunderstandings and incorrect reading of intelligence information).

Also, gaming can allow one to explore the potential impact of operational-level issues that often tend to be glossed over in policy-level work. For example:

1. National authorities often do not have some of the military options they assume they have, because the plans have not been laid, the procedures have not been developed, the information for carrying them out may not exist or there may be rather low-level constraints that preclude them;[14]

2. Some options might have dangerous consequences that would tend to go unrecognized without detailed simulations (e.g., some of the limited nuclear options that have been discussed from time to time look very different to the recipient team than they look to the team calling for them); and

3. Such games are well suited to exploring discontinuities. For example, the Control Team may march into the game room of the Blue Team and announce, 'Well . . . you no longer have any intelligence on Red's activities in Region X because Red has destroyed your intelligence collection system Y.' The Blue Team must then cope with a new situation.

The virtues of games, then, are considerable. There are also shortcomings, however, in that human games tend to focus on one or two scenarios, and may leave participants with lessons learned that would be the wrong lessons for other scenarios. The results also depend on who is playing the national leaders and so on. And, finally, human games are slow and expensive in terms of time and money. It is not feasible to collect the players one wants to have participate, except occasionally, and games with the people who *are* available may be highly unrealistic. As an aside, it has been comforting to note that players with greater proximity to policy and decision-making power have almost uniformly been the most reluctant to use nuclear weapons in games.

Artificial intelligence

For many years, computers were generally regarded as 'number crunchers', machines handy to have around after one had solved the problem except for tedious arithmetic. Although the original giants of computer science had no such illusions, recognizing that computers were essentially logic machines, the cold realities of computer performance limited their role until recently. There was, however, a research-level buildup of the concepts and techniques that are now so important in current practical work in expert systems and other fields of artificial intelligence (AI). A few are considered below.

Rule-based models for heuristic decision-making

As demonstrated in the pathbreaking work of Simon and colleagues in the 1950s and 1960s,[15] many of the most important problems faced by decision-makers must be addressed with heuristic reasoning rather than algorithmic approaches. Moreover, many decision-makers, and indeed experts in many domains, can solve problems essentially by applying rules. One semi-humorous definition of an expert is that an expert is someone who no longer needs to 'think' to solve problems—he merely 'knows what to do' (i.e., he has rules telling him which procedure to follow). This observation has, in recent times, led to the sub-field of artificial intelligence known as expert systems, and there now exists a vigorous technology concerned with building tools to make the building of such systems easier.[16]

Hierarchies

In some respects, an even more important contribution from AI research was the recognition of how important hierarchies are to complex organizations and any successful efforts to describe them. Related concepts involved 'organizational' or 'cybernetic' behaviour, and the heuristic reasoning mentioned above. Many of the issues were well discussed in the 1960s and 1970s by Simon,[17] Steinbruner[18] and others. It has now become commonplace, for example, to talk about bureaucratic models of national behaviour as the result of Allison's work in particular.[19]

What was lacking when many of these concepts emerged, however, was a mechanism for 'doing anything' with them other than writing a book or an essay. Although there were some primitive attempts at computer modelling to reflect the concepts, the attempts generally stayed at the level of academic toy problems. In large part this was because some of the people involved were not simulation-oriented, and in even larger part because the technology was not there to support it. Also, the capabilities were emerging in separate fields with only a modicum of cross-fertilization and integration. Roughly speaking, the author would argue that we have known for about 15 years how ill-suited standard methods of strategic analysis are in serving many of the requirements of decision-makers, but they have remained in use for want of something better for the practical down-to-earth business of preparing memoranda, studies, and books with graphs, tables and deductions allowing people to choose among options. It is now becoming possible to put the pieces together to do better.

IV. A first step towards a more ambitious methodology

In the preceding section simulation, gaming and artificial intelligence are discussed separately. Let us now recall some of the problems with traditional discussion of strategic policy issues: over-quantification, staticness, failure to represent national behaviour, and so on.[20] From this litany of problems coupled with the opportunities provided by technology and progress from the social sciences came the idea, in 1980, of developing an *automated war game*. RAND experimented with such ideas in 1980[21] and later began a substantial research programme[22] that is now bearing fruit. Figure 10.1 describes the automated war game in simple terms. The reader should remember that Rand is in the United States and that the game is therefore structured from the US viewpoint: there are entities called Red, Blue and Green agents representing

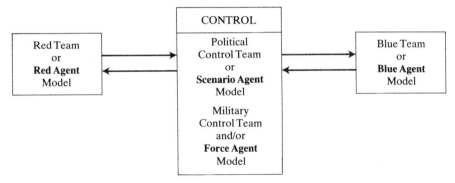

Red Agent: Makes decisions for the Soviet Union and the Warsaw Pact
Blue Agent: Makes decisions for the United States and NATO
Scenario Agent: Makes decisions for non-superpower countries
Force agent: Simulates results of combat and other military operations; also, controls simulation time

Figure 10.1. Structure of a game-structured simulation

the Soviet Union, the USA and third countries, respectively. The third countries represented by separate models within Green Agent fall into several categories: allies, neutrals and allies of the opponent. Each has its own national agenda as well as any coalition obligations.[23]

The essence of the system, then, is a structure[24] very much like the political–military war game pioneered by RAND in the 1950s.[25] There are, however, computer models to represent the various teams. Also, there is a computer model (Force Agent) to perform the Control Team services of keeping track of military forces, computing who wins the battles and at what cost, and so on. Force Agent is a large and sophisticated simulation model in its own right, one that is primarily quantitative in nature—although it also contains some rule-based models to provide operational and tactical-level command-control decisions such as how a theatre commander would allocate his forces on a daily basis.

The Red, Blue and Green Agents are rule-based models using heuristic rules to represent plausible behaviour patterns of nations. Their function is to issue the orders that Force Agent then carries out, and to communicate intentions and requests to the other decision models. The Red, Blue and Green Agents (decision models) can in principle be trivial or quite sophisticated.

As figure 10.2 indicates, the game proceeds with Red, Blue and Green taking turns making decisions, and Force Agent carrying out the associated orders. The decisions modelled by Red, Blue and Green range from top-level decisions about escalation or de-escalation down to the types of military decisions a theatre commander might make. As the second column of figure 10.2 indicates, each move of Red or Blue begins with an assessment of whether the current plan of action is proceeding acceptably. If it is, then the plan is

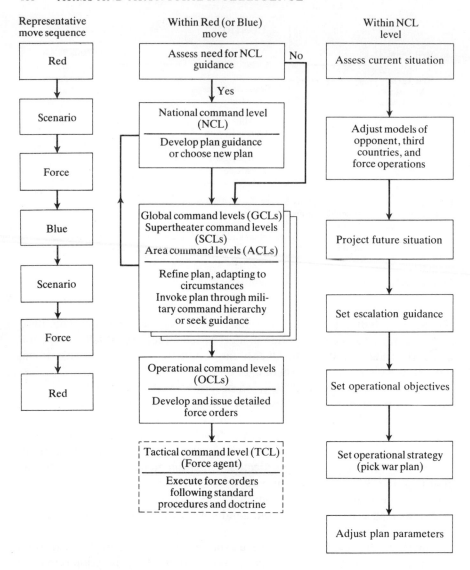

Figure 10.2. Conceptual model of the RSAC simulation

continued and a set of new orders may be issued accordingly (e.g., new orders allocating ground forces or air forces across both geography and mission). If it is not, then (as shown in the third column), a special sub-model awakes. This, the National Command Level model, has the opportunity to choose a new plan of action, to refine the current one, or to continue. The final decision on whether to go ahead with a new plan or not depends on the results of a 'look-ahead', which is a game within a game in which the Agent attempts to predict the consequences of choosing the plan using all the information it has. This look-ahead is technically identical to the basic game of which it is part except that the Agent's look-ahead is based on its *assumptions* about reality, which are often different from the game's reality. It is worth noting that human players in war games have seldom had the luxury of being able to make respectable look-aheads because they often have not been well organized, and because the Control Team couldn't handle the information requests in any case. In the context of the system described in figure 10.1, the look-ahead is merely another simulation.

Without going into details here, let it suffice to observe that the models that mimic the following of a plan and the models that mimic the choosing of a plan are both artificial intelligence models, but of very different types:

1. The military command level decisions are made by models intended to reflect organizational or cybernetic behaviour: the models follow a general course of action and will adapt to some extent to succeed in that course of action, but they will not attempt a global assessment of whether the grand plan is still reasonable. These models are hierarchical, with sub-models representing the various military commands down to the level of, for example, NATO's Commander-in-Chief of Central Command (CINCCENT).

2. The more strategic decisions determining which plan to follow in the first place, and when to change or substantially adapt a plan, are made by models intended to reflect strategic decision-making. These, the National Command Level (NCL) models, have a broader perspective and are explicitly tasked with worrying about national objectives, value trade-offs, and so on. These models also 'learn', as suggested by the third column of figure 10.2, in that they make their decisions on the basis of assumptions about the opponent and the environment, and they change those assumptions as the simulated crisis or war continues and they obtain empirical information.

NCL models have the potential to be far more realistic in their treatment of political–military decisions than previous models of which the author is aware because:[26] (*a*) their structure is patterned after real-world decision processes;[27] (*b*) the variables on which the decisions depend may be qualitative or quantitative; (*c*) the environment in which the models operate (the full game-structured simulation) encourages worrying about 'the right variables' (or, at least, the variables deemed important when human players engage in serious war games); and (*d*) the models recognize important processes such as characterizing the opponent, making reasonable assumptions about the laws of war, and—as noted above—adjusting those characterizations and assumptions in the light of further information.

NCL models are highly structured to reflect the hierarchical process of working from detailed information about the status of battle in one place or another up to strategic-level assessments that are part of decision-making on cosmic issues. The models are also variable to reflect uncertainty. Thus, there are alternative models of Red and Blue called Ivan 1, Ivan 2, . . . and so on. This is *essential* because uncertainties about national behaviour are fundamental, not merely a matter of insufficient research. Each NCL model has its own character, including a sense of grand strategy and guiding principles. These are determined by the analyst before writing the rules that constitute the model. The rules themselves are individually mundane, and become important and non-trivial only when taken as a whole. The rules are written in a special programming language called Rand-Abel, which is very easy to understand and modify, even for non-programmers.[28]

Figure 10.3 provides an example of one module of rules precisely as they appear in one of the computer programs. The issue is that Red is trying to characterize Blue in terms of several possible models of Blue (Blue 1, Blue 2, etc.), of which Blue 1 is the least likely to escalate. Note that the format is that of a decision table. The first rule in this table is if Current-situation is Eur-gen-conv *and* Warning-of-escalation is None *and* Time-since-D(Eur) is long *and* Presumed-opponent is unspecified. *Then* assume that the Presumed-opponent is Blue 1. By this we mean that if the current situation is a general conventional war in Europe, it has been some time (more than 10 days) since war began, and there is no warning of escalation, then Red will conclude that he is facing Blue 1.

The National Command Level models, of which only prototypes exist,[29] consist of thousands of rules, almost all of which are in decision tables such as this. These are not 'decision analysis models' in the sense of utility calculations: the rules are heuristic and the model as a whole is based on a human-like decision *process* rather than some mathematical calculation. The decisions can depend on 'hard' information (e.g., Warning-of-escalation would be evaluated as high if Blue had launched ICBMs and Red could see them) and 'soft' information (e.g., characterization of the opponent).

Decision Table
[-]

Current-situation	Warning-of-escalation	Time-since-D(Eur)	Presumed-opponent	/	Presumed-opponent
Eur-gen-conv	None	long	—		Blue
Eur-gen-conv	None	short	—		Presumed-opponent
Eur-gen-conv	Eur-nuc	—	Blue 1		Blue 3
Eur-gen-conv	Eur-nuc	—	>Blue 1		Presumed-opponent
Eur-gen-conv	>Eur-nuc	—	—		Blue 6

[long means greater than 10 days]
[the last rule reflects the nonpragmatic aspect of the case; hence, Blue is not Blue 5 and Ivan K now assumes the worst-case Blue 6]

[-]

Figure 10.3. Illustrative computer code in the National Command Level model

The use of models such as this to study issues of deterrence, escalation control, and war termination is still in its infancy, but even initial efforts have been intellectually rewarding—in large part because the quality of discussion on such matters is usually so hampered by phenomena such as people talking past each other, never quite agreeing on terms, and 'debating'. Consider, for example, the continuing problem of trying to understand when the opponent might escalate. Different experts on the subject may disagree violently on the matter. For example, those familiar with Soviet doctrine will observe that it emphasizes pre-emption, offensive actions generally, maintaining the initiative, and so on. On the other hand, those familiar primarily with Western concepts of deterrence may talk about bargaining, escalation ladders, and so on. At first glance, these views are irreconcilable. However, by attempting to formalize the issues through modelling some of the disputes can be resolved and the realm of disagreement narrowed to sharply posed issues. A major factor here is that modelling forces specificity of context: one cannot talk about what the opponent might or might not do as a generality, but rather must talk about what the opponent might do in a very specific situation (such as might arise in a human war game or, in our work, in a game-structured simulation). The result is models of Red that will follow doctrine scrupulously under some circumstances and not at all under others. Similarly, our Blue models have complex rather than stereotyped behaviour.

Another conclusion quickly reached upon attempting to write serious National Command Level models addressing DEWT issues is that the 'real' variables that would affect decision are indeed quite different from those emphasized in traditional strategic analysis. It is not only possible but natural to have the decisions depend on the types of issue mentioned at the outset of the paper as having been chronically neglected (e.g., command-control problems, alliance problems, and assessments of the various actors). It remains to be seen how far this methodology can be pushed, but initial work suggests that it has a rich potential for allowing us to represent in a single framework the wisdom of both the hard and soft sciences on the cosmic issues of strategic policy.

Notes and references

[1] Portions of this paper appear in Davis, P. K., 'A new analytic technique for the study of deterrence, escalation control, and war termination', in Cimbala, S. J. (ed.), *Artificial Intelligence and National Security* (Lexington Books: Lexington, MA, 1986), and Davis, P. K., Bankes, S. C. and Kahan, J. P., *A New Methodology for Modeling National Command Level Decisionmaking in War Games and Simulations*, The Rand Corporation, R-3290-NA, 1986.

[2] Brodie, B., *Strategy in the Missile Age* (Princeton University Press: New Jersey, 1959).

[3] Schelling, T. C., *Arms and Influence* (Yale University Press: New Haven, CT), 1966.

[4] Kahn, H., *On Escalation: Metaphors and Scenarios* (Penguin Books: Baltimore, MD, 1968).

[5] Wohlstetter, A., 'The delicate balance of terror', *Foreign Affairs*, vol. 37, no. 2 (Jan. 1959), pp. 211–56.

[6] Ikle, F. C., *Every War Must End* (Columbia University Press: New York, 1971).

[7] See Davis, Bankes and Kahan (note 1), and Davis, P. K. and Stan, P. J. E., *Concepts and Models of Escalation*, The Rand Corporation, R-3235, May 1984.

[8] Prigogine, I. and Stengers, I., *Order Out of Chaos: Man's Dialogue with Nature* (Bantam Books: New York, 1984).

[9] Keeney, R. L. and Raiffa, H., *Decisions With Multiple Objectives: Preferences and Value Tradeoffs* (John Wiley and Sons: New York, 1976).

[10] Bracken, P., *The Command and Control of Nuclear Forces* (Yale University Press: New Haven, CT, 1983); see also Blair, B., *Strategic Command and Control* (Brookings Institution: Washington, DC, 1985).

[11] Forrester, J. W., *Urban Dynamics* (MIT Press: Cambridge, MA, 1969).

[12] Other types of stimulation modelling have been used extensively in system-level strategic analysis for some time—e.g. for design and evaluation of ballisic missiles, penetration aids, defences, and their interactions. It is their use in policy analysis that has lagged.

[13] Brewer, G. and Shubik, M., *The War Games: A Critique of Military Problem Solving* (Harvard University Press: Cambridge, MA, 1979).

[14] See note 10.

[15] Simon, H. A., *Sciences of the Artificial* (MIT Press: Cambridge, MA, 1980).

[16] Hayes-Roth, F., Waterman, D. A. and Lenat, D. B. (eds), *Building Expert Systems* (Addison-Wesley: Reading, MA, 1983).

[17] See note 15.

[18] Steinbruner, J. D., *The Cybernetic Theory of Decision: New Dimensions of Political Analysis* (Princeton University Press: New Jersey, 1974).

[19] Allison, G. T., *Essence of Decision* (Little, Brown & Co.: Boston, MA, 1971).

[20] Marshall, A. W., 'A program to improve analytic methods related to strategic forces', *Policy Sciences*, vol. 15, no. 1 (Nov. 1982).

[21] Graubard, M. H. and Builder, C. H., *Rand's Strategic Assessment Center: An Overview of the Concept*, The Rand Corporation, N-1583-DNA, Sep. 1980.

[22] Davis, P. K. and Winnefeld, J. A., *The Rand Strategy Assessment Center*, The Rand Corporation, R-2945-DNA, 1983.

[23] See Schwabe, W., *Strategic Analysis As Though Nonsuperpowers Matter*, The Rand Corporation, N-1997-/DNA (a doctoral dissertation at the Rand Graduate School), 1983.

[24] See Davis, Bankes and Kahan (note 1).

[25] See discussion in Brewer and Shubik (note 13) of work by Goldhamer and Spier, among others.

[26] See also Davis and Stan (note 7), Davis (note 1), Bankes and Kahan (note 1) and Davis, P. K., Stan, P. J. E. and Bennett, B. W., *Automated War Gaming as a Technique for Exploring Strategic Command and Control Issues*, The Rand Corporation, N-2044-NA, Nov. 1983.

[27] See, for example, Janis, I. L. and Mann, L., *Decision Making* (The Free Press: New York, 1977).

[28] Shapiro, N. Z., Hall, H. E., Anderson, R. H. and Lacasse, M., *The Rand-Abel Programming Language: History, Rationale, and Design*, The Rand Corporation, N-2367-NA, Oct. 1985.

[29] See Davis, Bankes and Kahan (note 1).

Chapter 11. Verification and stability: a game-theoretic analysis

STEVEN J. BRAMS AND D. MARC KILGOUR

I. Introduction

Arms control between the superpowers seems at a dead end unless the provisions of any treaty they negotiate can be verified. In a recent assessment, one expert argued that 'verification has become the most important standard against which arms control agreements—both past and prospective—are measured',[1] and then went on to point out that:

five years ago . . . verification was a contentious issue in the arms control debate, but it was not the decisive one. Today, this situation has changed, and verification stands as the litmus test by which arms control proposals are assessed.[2]

Another expert expressed a similar viewpoint about the primacy of verification:

The ability of the United States to verify Soviet compliance with the provisions of arms control agreements . . . has been taken by many as a fundamental requirement for establishing that a treaty is in the interests of the United States.[3]

The formal analysis of verification, based on strategic models of rational choice, has a curious history. Twenty years ago, such analysts as Maschler[4] and Rapoport[5] developed rather elaborate game-theoretic models to analyse the strategic implications of different inspection procedures. Since then, except for some scattered and mostly unpublished reports, game theory and other formal tools of rational strategic analysis were not again applied to arms control issues until recently. Some of this work is reviewed in the next section but first the scope of this chapter and its main results are outlined.

First, as a basic model of verification a variable-sum game between an inspector and an inspectee is proposed. It is different from games that have previously been analysed in the literature. For reasons given below, it is believed that this game better captures the crucial features of the strategic conflict between the superpowers over the issue of verification.

Endowing the inspector with an ability to detect the choices of the inspectee, but only imperfectly, analysis is first made of the pay-offs for the players from the inspector's inducing the inspectee to respond to its announced strategy. Inducement may be thought of as a device for one player's leading the other to do something, or deterring it from doing something, in such a way as ultimately to benefit itself.

Next all Nash[6] equilibria in this game are systematically examined. The dependence of equilibria and inducement outcomes on the inspector's detection probabilities—both to ascertain compliance and uncover non-compliance—are shown.

The two solution concepts of inducement and equilibrium are then compared, and the applicability of each to practical verification issues discussed. The general goal is to provide a theoretical framework for viewing such issues, indicating what kinds of strategies are appropriate for achieving outcomes that help to dispel distrust and stabilize co-operation.

Trust and stability are closely linked: adversaries can afford to trust each other if an outcome is stable—that is, a unilateral deviation will hurt the deviator—whether the players' choices are simultaneous (Nash equilibrium) or sequential (inducement outcomes). Thus, if the inspectee is robbed of the incentive to cheat, and the inspector to hide or manipulate the information it collects, trust will be fostered.

Moreover, this is trust based not on mere blind faith but rather on calculations of advantage and disadvantage. In the context of arms control, conditions are derived under which it is more costly to violate than not violate the provisions of a treaty when the players choose strategies that are in some sense stable.

Verification may therefore be connected to stability and ultimately to trust. In fact, the connection is two-sided: verification can give one good reason both to trust an opponent and to be trustworthy to him.

This is not to say that present technologies for verifying adherence to arms control agreements are adequate for all weapon systems. Immense practical problems exist in detecting the deployment of certain strategic weapons, such as cruise missiles. An important by-product of this game-theoretic analysis—in addition to clarifying the qualitative notions of stability and inducement—is the demonstration that there are quantitative thresholds, as measured by the level of detector reliability, sufficient to ensure different kinds of stable and co-operative outcomes.

II. The Verification Game

A number of different games have been proposed as models of the strategic situation facing a possible violator of an arms control agreement—the Inspectee—and a detector who signals possible violations—the Inspector. Before reviewing these, the authors posit their own game the (Basic) Verification Game (VG)—and some comparisons are drawn between this game and others that have been proposed as models of verification.[7]

In VG it is assumed that Inspectee may either comply with (C) or violate (i.e., not comply with—$\bar{C}$) an arms control agreement, but that Inspectee always *claims* that he chose C. Inspector may either accept (A) or challenge (i.e., not accept—$\bar{A}$) Inspectee's stated compliance. (Mixed strategies are used below to model the possibility that Inspectee may choose any *level* of compliance or non-compliance, and Inspector any *degree* of acceptance or non-acceptance, but this complication is not necessary to describe the basic game.)

In the basic version of VG, each player has two strategies that lead to four possible outcomes at the intersection of each pair of strategy choices, as shown in figure 11.1. It is assumed that the pay-offs to the players are normalized, so that the best outcomes for Inspectee (ROW) and Inspector (COL) have utilities $r_4=1$ and $c_4=1$, and the worst have utilities $r_1=0$ and $c_1=0$, respectively. The players' next-best pay-offs are r_3 and c_3, and next-worst r_2 and c_2. In summary, the pay-offs satisfy

$$0=r_1<r_2<r_3<r_4=1$$
$$0=c_1<c_2<c_3<c_4=1$$

VG is defined by the pay-offs at the four outcomes shown in figure 11.1. These pay-offs are assumed to be cardinal utilities and are indicated by ordered pairs (r_i, c_j). For example, $(r_3, 1)$ at the upper-left outcome of figure 11.1, associated with the strategies C and A, gives ROW its next-best pay-off and COL its best pay-off.

		Inspector (COL)		
		Accept (A)	Challenge/ Don't accept ($\bar{A}$)	
Inspectee (ROW)	Comply (C)	$(r_3, 1)$	$(0, c_3)$	(s)
	Violate/ Don't comply ($\bar{C}$)	$(1, 0)$	(r_2, c_2)	$(1-s)$
		(u)	$(1-u)$	

Key: (r_i, c_j) = (pay-off to ROW, pay-off to COL).
Normalization: $0=r_1<r_2<r_3<r_4=1$
$0=c_1<c_2<c_3<c_4=1$

Figure 11.1. The basic Verification Game (VG)

The pay-off rankings in VG can be characterized by the primary (1) and secondary (2) goals of each player:

ROW (Inspectee)
1. Prefers COL *accept* its claimed compliance (two best outcomes associated with A, two worst with $\bar{A}$).
2. Prefers to *violate* agreement (whether COL chooses A or $\bar{A}$, ROW prefers $\bar{C}$).

COL (Inspector)
1. Prefers ROW *comply* (two best outcomes associated with C, two worst with $\bar{C}$).

2. Prefers policy of *tit-for-tat* (if ROW chooses C, COL prefers A; if ROW chooses $\bar{C}$, COL prefers $\bar{A}$).

These goals determine a lexicographic order,[8] whereby the primary goal distinguishes the two best from the two worst outcomes for each player; given this distinction, the secondary goal orders the two best, on the one hand, and the two worst, on the other, for each player.

Thus for ROW, its primary goal says that its two best outcomes are in the A column of figure 11.1, and its secondary goal says that between these two outcomes it prefers the one associated with $\bar{C}$. Therefore, its pay-off from strategy pair $\bar{C}A$ is $r_4=1$, and its pay-off from strategy pair CA is r_3; similarly, between its two worst pay-offs in the $\bar{A}$ column (as implied by its primary goal) its secondary goal says that $\bar{C}\bar{A}$ yields r_2 and $C\bar{A}$ yields $r_1=0$.

The primary goals of each player would certainly appear plausible: ROW would want its *claimed* compliance accepted, and COL would want *actual* compliance. Also, ROW's secondary goal, if morally dubious, is probably realistic in many situations. For if violating an agreement were not profitable, then there would be no reason for ROW not to comply, making for a trivial game with a mutually best (1, 1) outcome.

Similarly, COL's secondary goal seems eminently defensible. First, given ROW chooses C, COL's preference for A is reasonable, for why should COL not co-operate by choosing A when ROW co-operates by choosing C? The second part of this tit-for-tat goal seems equally reasonable—that is, for COL to be most hurt when it accepts a violation.

If the reasonableness of the goals of the players is not at issue, the question of which goal is primary and which is secondary may be more controversial. Conceivably, ROW may prefer violating an agreement over having it accepted by COL, and COL may prefer a policy of tit-for-tat over ROW's compliance. A reversal in the priority of primary and secondary goals by one player would give rise to two new games, and a reversal by both would generate a third.

Different variable-sum games have been analysed by other theorists. For example, Rapoport[9] applied a number of different solution concepts to a game between an inspector and an evader, in which the primary and secondary goals of the inspector duplicated those in VG. However, the evader had a primary goal of deceiving the inspector—not simply having its claimed compliance accepted; its secondary goal was the same as our inspectee's primary goal. Maschler[10] proposed more elaborate inspection games, involving chance and an inspector who could decide whether or not to investigate a suspicious event. In his model, Maschler assumed that the inspector could announce and commit itself to a mixed strategy, against which the potential violator would maximize its expected pay-off.

Brams and Davis[11] and Brams[12] used this idea of inducement, which was originally proposed by von Stackelberg[13] to study price leadership, and applied it to a 'truth game' to model superpower verification. In this game, which reverses the priority of the primary and secondary goals of the inspector but not the evader in Rapoport's[14] game, they investigated both 'inducement' and 'guarantee' strategies. More specifically, they analysed the benefits that not only the inspector but also the inspectee could realize by inducing the other player to respond to it, comparing these with the pay-offs that the players could guarantee for themselves whatever the opponent's choice (analogous to minimax/maximin strategies in two-person constant-sum games). More recently, Brams and Kilgour[15] studied the Nash-equilibrium strategies of the players in Rapoport's inspector-evader game and the truth game.

Fichtner[16] compared various game-theoretic approaches to the inspection and verification of arms control agreements; as well, he applied similar solution concepts to auditing, consumer protection and nuclear safeguards. Avenhaus's[17] treatment of the latter issue, in particular, makes significant use of game theory; among other things, he finds optimal strategies for an inspector, with goals the same as those in VG, to induce compliance by an inspectee, whose primary but not secondary goal is the same as that in VG.

There is probably no 'best' game to model all aspects of verification. VG is justified above in terms of the primary and secondary goals of the players, but it is worth noting why the VG ranking seems especially plausible in the case of the superpowers. Consider each of the four outcomes in turn, and assume that one superpower (ROW) contemplates cheating on an arms control treaty and the other (COL), which monitors its compliance, can challenge possible violation:

$\overline{CA}$—(1,0): A successful violation of a treaty, giving one side a substantial edge in the arms race, would certainly seem the best outcome for ROW and the worst for COL.

CA—(r_3,1): This is the best outcome for COL, for it validates the treaty, but it is definitely inferior for ROW because an unchallenged violation could give ROW the edge mentioned above.

$C\overline{A}$—(0,c_3): COL gets compliance but, without at first recognizing it, creates some distress for itself by erroneously challenging ROW; for ROW, on the other hand, a false charge of cheating is its worst outcome, undermining the benefits of the treaty for no gain.

$\overline{CA}$—(r_2,c_2): A rather unsatisfactory outcome for both sides, because COL's justified challenge of violations underscores the treaty's fragile status, perhaps leading to its abrogation.

One might contend that $\overline{CA}$ would be higher than next-worst in COL's preferences, for challenging non-compliance is certainly in COL's interest. It is believed, however, that COL would prefer compliance at $C\overline{A}$ to non-compliance at $\overline{CA}$, even though compliance at the former outcome includes the embarrassment of a false charge.

The reason for this preference is that the damage done by an unjustified accusation can more easily be undone than that of an actual violation, even detected and challenged. After 40 years of conflict, both superpowers almost certainly would prefer to have their adversary adhere to SALT and other arms control treaties, even at the cost of occasionally making unsubstantiated charges, than to catch their adversary in a lie and challenge real violations.

Whether c_2 and c_3 are interchanged in VG, however, makes no difference for the rational strategy choice of ROW, which is to choose $\overline{C}$. For this strategy is *dominant*: whether COL chooses A or $\overline{A}$, $\overline{C}$ is better than C for ROW. In a game of complete information, COL would know that ROW has an unconditionally best strategy choice, and presuming ROW would choose it, COL could do no better than choose $\overline{A}$, leading to (r_2, c_2).

This is the unique *Nash equilibrium* in VG: once at this outcome, neither player would have an incentive to depart from it unilaterally because it would do worse if it did. Unfortunately for both players, however, this outcome is *Pareto-inferior*: it is worse for both players than $(r_3, 1)$. Yet the latter outcome is not in equilibrium because ROW has an incentive to depart from it to $(1, 0)$, its best outcome.

The Pareto-inferior rational solution to this game can be circumvented if, assuming ROW chooses its strategy first, COL has perfect information about ROW's choice and ROW knows this. In this case, it is easy to show that COL would have a dominant strategy of tit-for-tat—choose A if ROW chooses C, and $\overline{A}$ if ROW chooses $\overline{C}$—and ROW, anticipating this dominant-strategy choice, would choose C, resulting in $(r_3, 1)$, a Nash equilibrium in the resulting 2×4 game.

But, of course, the superpowers in general will have only imperfect information about each other's choices when each plays the role of Inspector in VG. How, then, can they use their imperfect detection equipment to choose optimally? The next section proposes two different notions of 'optimal' and analyses how rational strategies in each case depend on the quality of the detection equipment in a more realistic version of VG.

III. Inducement in the verification game with detection

It is assumed that COL is equipped with a detector and has the option of consulting it before choosing A or $\overline{A}$. COL's detector is characterized by parameters x and y, which are conditional probabilities that describe its reliability and are assumed to be known by both players:

x=Pr{detector signals violation | ROW chose $\overline{C}$}

y=Pr{detector signals no violation | ROW chose C}

Note that $0 \leq x, y \leq 1$. The detector is perfect when $x=y=1$ and would seem worthless when x and/or y are near or at 0.[18]

It is assumed that ROW chooses its strategy before COL. ROW has two *pure*, or single, strategies, C and $\overline{C}$, so its mixed strategy, which is simply a probability distribution over its set of pure strategies, can be represented by a single probability s:

Pr{ROW chooses C}=s.

Because COL has a detector, its choices are more complicated. Besides its two pure strategies, A and $\overline{A}$, which are treated as *unconditional choices* (i.e., made without consulting the detector), it has a third pure strategy of consulting its detector (D). If COL chooses D, it is assumed it will follow a policy of tit-for-tat by picking $\overline{A}$ if the detector signals a violation and A otherwise. Altogether, COL's mixed strategy is represented by two probabilities, t and u, where

Pr{COL chooses D}=t

Pr{COL chooses A}=u

Of course, Pr{COL chooses $\overline{A}$}=1−t−u, just as Pr{ROW chooses $\overline{C}$}=1−s. Because the mixed strategies are probabilities,

$0 \leq s \leq 1$; $0 \leq u \leq 1$; $0 \leq t \leq 1$; $u+t \leq 1$;

the latter sum is the probability that COL chooses either A or D—that is, does not choose $\overline{A}$.[19]

The probabilities of ROW's choices C and $\overline{C}$ (s and 1−s), and the unconditional probabilities of COL's choices A and $\overline{A}$ (u and 1−u), are indicated in parentheses to the right and below these strategies in figure 11.1. Combining these probabilities with the probability, t, of COL's consulting its detector (D), and then choosing either A or $\overline{A}$, produces the probabilities that each of the four possible outcomes of the game will occur:

CA: s(u+ty)

C$\overline{A}$: s(1−u−ty)

$\overline{C}$A: (1−s)(t+u−tx)

$\overline{CA}$: (1−s)(1−t−u+tx)

For example, CA is chosen when ROW chooses C (with probability s) and COL chooses either A unconditionally (with probability u) or consults its detector (with probability t) and, detecting no violation (with probability y), chooses A.

These outcome probabilities may now be combined with the pay-offs that the players obtain at the outcomes (see figure 11.1) to give the players' expected pay-offs (Es) in VG. To simplify notation, we shall henceforth distinguish the Es by subscripts R (for ROW) and C (for COL):

$E_R(s;t,u) = s(u+ty)r_3 + (1-s)(t+u-tx) + (1-s)(1-t-u+tx)r_2$

$E_C(t,u;s) = s(u+ty) + s(1-u-ty)c_3 + (1-s)(1-t-u+tx)c_3$

In appendix 11A, it is shown that the maximin strategies for ROW and COL—that is, the strategies that maximize their minimum expected pay-offs, whatever the opponent does—are s=0 and u=t=0, respectively. These strategies, which are for ROW always to choose $\overline{C}$ and for COL always to choose $\overline{A}$ unconditionally (and never consult its detector), give the players maximin values of r_2 and c_2, respectively. Recall that these are the pay-offs to the players from choosing their Nash-equilibrium strategies in the figure 11.1 game (without the possibility of detection by COL).

The Nash equilibria that arise when COL has an imperfect detector that it can use to try to discern ROW's (prior) strategy choice are investigated below. They can be different from the unique Nash equilibrium in the figure 11.1 game without detection.

First, however, it is investigated how COL, by announcing and committing itself to a mixed strategy (u, t), can induce ROW to respond to this strategy in such a way that COL maximally benefits. As shown, COL's optimal inducement strategy depends on its detector's having detection probabilities, x and y, above a certain minimum. It is assumed, for this exposition, that both x<1 and y<1; this assumption is dropped in appendix 11A.

Because ROW has only two pure strategies, COL can induce either one or the other by arranging that it be ROW's unique best response. It cannot induce a mixture since no mixed strategy could constitute ROW's best response unless all do, which is to say that no strategy, pure or mixed, is better than any other.

It is shown in the appendix that $s \dot= 0$ (always choose $\overline{C}$) is a best response for ROW to any strategy of COL if

$$x(1-r_2)+yr_3 \leq 1.$$

But then COL cannot receive more than c_2, whereas if s=1 COL cannot receive less than $c_3 > c_2$ (see figure 11.1).

Obviously COL would prefer to induce s=1, and he can do so provided that

$$x(1-r_2)+yr_3 > 1. \tag{1}$$

From appendix 11A it is seen that, under this condition, COL induces s=1, and benefits maximally, when it chooses

$$u = \left[\frac{x(1-r_2)-1+yr_3}{x(1-r_2)-r_3+yr_3}\right]^- ; \quad t=1-u = \left[\frac{1-r_3}{x(1-r_2)-r_3+yr_3}\right]^+, \tag{2}$$

where the '−' and '+' superscripts indicate values slightly less and greater, respectively, than those given in the brackets. This strategy makes s=1 ROW's unique best response, and among all strategies that do so this one maximizes COL's expected pay-off.

What are the benefits to COL of its optimal inducement strategy? It turns out that this strategy yields COL

$$E_C^* = E_C(u,t;1) = \left[\frac{x(1-r_2)-(1-y)(1-c_3+c_3r_3)}{x(1-r_2)-(1-y)r_3}\right]^-,$$

and $c_3 < E_C^* < 1$. Thus, COL receives strictly more than its next-best pay-off of c_3; it receives somewhat less than its best pay-off of 1 because of detector unreliability.

For ROW the benefits are not so great—in terms of the comparative rankings of the players—but neither are they terrible:

$$E_R^* = E_R(1;u,t) = \left[r_3 \left(\frac{x(1-r_2)-1+y}{x(1-r_2)-r_3+yr_3} \right) \right]^-,$$

and $r_2 < E_R^* < r_3$. In other words, ROW receives strictly more than its maximin value of r_2, but strictly less than its pay-off at the 'co-operative' outcome $(r_3, 1)$, when $s = u = 1$ and $t = 0$.

Indeed, $(r_3, 1)$ gives a better pay-off to *both* players than what COL can induce, so inducement is Pareto-inferior to the pure strategy outcome $(r_3, 1)$. Moreover, even those inferior inducement pay-offs are unattainable unless the conditional detection probabilities, x and y, are sufficiently high that inequality (1) is satisfied. Given that (1) is satisfied, COL can induce either $s=0$ or $s=1$; by choosing u and t according to (2), it will induce (E_R^*, E_C^*), which is certainly better for both players than (r_2, c_2) (obtained by optimally inducing $s=0$), even if it is Pareto-inferior to $(r_3, 1)$.

It might be thought that ROW could turn the tables on COL and induce COL to respond to its own mixed strategy. However, ROW would have no reason to announce that it might ever choose $\overline{C}$—violate the treaty—because this could only steer COL toward the choice of $\overline{A}$, which leads to ROW's two worst outcomes.

In the next section it is assumed that COL is not able to seize the initiative and announce its optimal inducement strategy to evoke a best response from ROW. Instead it is asked whether COL, again with only an imperfect detector, and ROW can simultaneously (or at least in ignorance of each other) choose strategies such that neither would have an incentive to depart from its choice unilaterally.

IV. Nash-equilibrium strategies

Assume that ROW does not respond to COL's prior choice—as under optimal inducement by COL—perhaps because COL is unable to make a credible commitment to a particular mixed strategy. Rather, suppose that both players, knowing that COL can imperfectly detect ROW's prior choice, act in light of this knowledge (i.e., of the conditional probabilities, x and y) and the pay-offs of VG shown in figure 11.1.

Under the assumption that $x>0$ and $y>0$ (i.e., the detector has non-zero probabilities of being correct), we derive in the appendix all Nash equilibria in VG with detection. These equilibria can be classified into five distinct groupings:

I. *Co-operative equilibrium*, with pay-offs of $(r_3, 1)$. Since this equilibrium requires $y=1$ (perfect detection of compliance by COL) it is unrealistic and is not considered further.

II. *Non-co-operative equilibria*, with pay-offs of (r_2, c_2). These equilibria can always (i.e., for any values of x and y) be achieved by COL's choosing $\overline{A}$ for certain and ROW's choosing $\overline{C}$ for certain. If $x=1$, COL may instead consult his detector some of the time without changing the equilibrium.

In any event, these equilibria give the players only their maximin values and are dominated by all other equilibria whenever other equilibria exist (more on this point below).

III. *Constant-detection equilibrium*, with pay-offs

$$(yr_3, xc_2+s[c_3+y(1-c_2)-xc_2]),$$

where $x(1-r_2)+yr_3=1$. COL always consults its detector at this equilibrium and so follows a policy of tit-for-tat. This equilibrium is not very important because it occurs very rarely—only when the values of x and y precisely satisfy a linear equation.

IV. *Never-accept (unconditionally) equilibrium*, with pay-offs

$$\left(\frac{yr_2r_3}{yr_3-(1-x)(1-r_2)}, \frac{yc_2(1-c_3)+(1-x)c_2c_3}{y(1-c_3)+(1-x)c_2}\right),$$

where $x(1-r_2)+yr_3>1$. COL either consults its detector or chooses $\overline{A}$ but never chooses A; because this equilibrium is dominated by v below (i.e., v is better for both players than IV), it is not considered further.

V. *Never-challenge (unconditionally) equilibrium*, with pay-offs

$$\left(\frac{x(1-r_2)r_3-(1-y)r_3}{x(1-r_2)-(1-y)r_3}, \frac{xc_2}{(1-y)(1-c_3)+xc_2}\right),$$

where $x(1-r_2)+yr_3>1$. COL either consults its detector or chooses A but never chooses $\overline{A}$.

To summarize, I is summarized as unrealistic because perfect detection is unattainable, III as unimportant because it almost never occurs, and IV because it is dominated by v. It is also shown in the appendix that II is dominated by v (when v occurs); only the $\overline{CA}$ form of II is considered here for the x=1 variant requires perfect detection.

This leaves II and V. If

$$x(1-r_2)+yr_3<1, \tag{3}$$

II is the only equilibrium. The unshaded region below and to the left of the line shown in figure 11.2 satisfies inequality (3).

Thus, if the detection probabilities, x and y, are so low as to satisfy (3), only strategies associated with the non-cooperative equilibrium are stable in VG. Formally, these strategies are

s=0; u=0, t=0,

where x<1.

The situation is a little better for the players if equality holds in (3), for equilibrium III, which dominates II, then exists. But this occurs only *on* the line $x(1-r_2)+yr_3=1$, shown in figure 11.2.

The picture is still brighter if inequality (3) is reversed, which is in fact inequality (1) in section III above. If x and y are sufficiently high that $x(1-r_2)+yr_3>1$, then ROW and COL can obtain the pay-offs given by equilibrium V. Of course, equilibria II and IV are also available in this region, which lies above and to the right of the line $x(1-r_2)+yr_3=1$, shown in figure 11.2. Presumably, however, the players will choose the dominant equilibrium in this region—that which is better for both players.

Consider the region in which equilibrium V exists and is dominant. This region, which is defined by inequality (1), is always the region in which COL can induce ROW to choose s=1 (i.e., always comply), as shown in section III.

A comparison of COL's *pay-off* at equilibrium V with what it can obtain through optimal inducement is instructive. As shown in appendix 11A COL does better under optimal inducement, obtaining E_C^*, than it does at equilibrium V; ROW does marginally worse.

In fact, COL's optimal inducement strategy given by (2) is almost identical to its strategy at equilibrium V (see appendix 11A for details). In either case, COL sometimes consults its detector, sometimes chooses A, but never chooses $\overline{A}$ without first consulting its detector. ROW's equilibrium V strategy is mixed (see appendix 11A), but of course its best response to COL's optimal inducement is pure (C).

As the detection probabilities, x and y, approach 1—allowing for perfect detection—the pay-offs to the players approach $(r_3, 1)$ both under optimal inducement and at equilibrium V. In other words, the players can come closer and closer to the pay-offs of the co-operative equilibrium (I) as detection improves, which is a hopeful sign in VG. Moreover, COL need never resort to

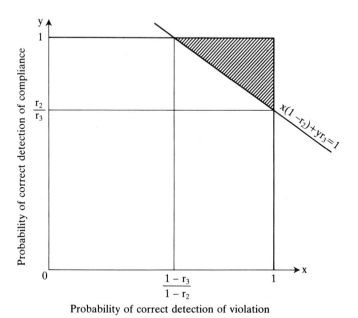

Probability of correct detection of violation

Figure 11.2. The region in which equilibrium V dominates all other Nash equilibria and inducement is possible. The region in question is shaded.

choosing $\overline{A}$ unconditionally—before consulting its detector—as long as inequality (1) is satisfied.

Only below the line shown in figure 11.2, where inequality (3) is satisfied and detection is relatively poor, do the players have no equilibrium strategies other than always to choose $\overline{C}$ and $\overline{A}$. Above this line, it is advantageous for COL sometimes to consult its detector and sometimes to accept unconditionally. This mixed strategy evokes greater compliance from ROW, which renders the outcome at equilibrium v stable and yields higher pay-offs to both players than are available at any other Nash equilibrium. However, whenever equilibrium v is available, it is always in COL's interest to attempt optimal inducement, for this approach would substantially improve COL's expected pay-off without any significant effect on ROW's. These ideas are illustrated by a numerical example: if $r_2=0.2$, $r_3=0.7$, $c_2=0.4$, $c_3=0.7$, and $x=y=0.8$, then (1) holds, and the players have equilibria II, IV, and V available, paying (0.2, 0.4), (0.28, 0.48), and (0.62, 0.84), respectively. At equilibrium v, ROW complies 84 per cent of the time; COL consults its detector 60 per cent of the time, and accepts unconditionally otherwise. For optimal inducement, COL credibly commits itself to consulting its detector slightly more than 60 per cent of the time and accepting unconditionally otherwise. ROW's best response is to increase compliance to 100 per cent, and the pay-offs become $(0.62^-, 0.96^-)$; hence, COL's gain of almost 0.12 is achieved at essentially no cost to ROW.

V. Conclusions

There are both auspicious and inauspicious implications of this analysis of the Verification Game with detection. Inauspiciously, the detection probabilities must be above threshold values before the detector can be helpful at all, or inducement by Inspector is worth trying. Indeed, below these values the detector is worse than useless and should never be consulted. Instead, the players' strategies of always violating and always challenging are the only strategies in equilibrium.

Auspiciously, though the players' non-cooperative strategies remain in equilibrium above the threshold values, this equilibrium is dominated by more co-operative equilibria that yield higher pay-offs for both players. These equilibria involve mixed strategies, whereby Inspector sometimes relies on its detector, sometimes accepts (unconditionally) Inspectee's claimed compliance, but never challenges unconditionally. Inspectee's equilibrium strategy is also mixed, but between compliance and non-compliance, with the probability of compliance rising as Inspector's detector improves.

By announcing a mixed strategy to which Inspectee responds, Inspector can induce a pure-strategy response by Inspectee that raises Inspector's pay-off above its greatest Nash-equilibrium value (in the shaded region in figure 11.2). Inspectee is slightly hurt in this region when it is induced to comply.

Whatever the differences between the inducement and equilibrium pay-offs, the key to 'solving' the Verification Game lies not so much in the solution concept selected but in being the favorable region (shaded in figure 11.2), which triggers both the possibility of inducement and Nash equilibria that are Pareto-superior to the non-cooperative equilibrium. Inducement probably makes most sense if there is a third party or neutral inspector, which can credibly announce acceptance conditions to which the parties to a conflict will respond. On the other hand, in a two-person game between adversaries such as the superpowers—wherein each adversary plays the role of both Inspector and Inspectee—the Nash-equilibrium strategies may be more sensible because of their symmetry with respect to the players.

It is difficult to say whether the verification equipment of the superpowers is sufficiently good to place them in the favourable region. Nevertheless, the nuclear arms control agreements reached so far, though limited in scope, argue against unbridled pessimism.

Verification of the most significant agreements that the United States and the Soviet Union have achieved, beginning with the Limited Nuclear Test Ban Treaty prohibiting nuclear weapon tests in the atmosphere, outer space, and under water in 1963, has been based on 'national technical means' (NTM). The use of NTM means that each superpower relies primarily on its own surveillance of the other side's activities, using reconnaissance satellites and the like. The importance of NTM since the early 1960s should not be underestimated; in fact, without NTM, there would be no Verification Game.[20]

Perhaps we cannot be so sanguine about new and future weapon systems, including those, like cruise missiles, that are highly mobile or those, like anti-satellite weapons, that are hard to detect for technical reasons. Yet, because it is in both sides' interest—whether in the role of Inspector or Inspectee—to be in the favorable region, this analysis suggests that each side's apparent obsession to conceal as much as possible is probably ill-founded.

It is worth stressing that stability, except of the non-cooperative kind, is impossible unless detection thresholds are reached and surpassed. In the absence of such stability achieved by equilibrium or inducement, verification will surely fail, for the rational strategies of players—to violate and challenge—will undermine it. The authors believe that the primary goal of verification must be to develop procedures for exceeding the thresholds at which cooperative behaviour is stabilized. The players have a mutual interest in being both more compliant and more willing to develop and allow the use of better detection methods.

Notes and references

[1] Potter, W. C., 'Introduction', in Potter, W. C. (ed.), *Verification and Arms Control* (Lexington Books: Lexington, MA, 1985), p. 1.

[2] Potter (note 1), p. 4.

[3] Richelson, J., 'Technical collection and arms control', in Potter (note 1), p. 202.

[4] Maschler, M., 'A price leadership method for solving the inspector's non-constant-sum game', *Naval Research Logistics Quarterly*, vol. 13 (1966), pp. 11–33; and Maschler, M., 'The inspector's non-constant-sum game: its dependence on a system of detectors', *Naval Research Logistics Quarterly*, vol. 14 (1967), pp. 275–90.

[5] Rapoport, A., *Two-Person Game Theory: The Essential Ideas* (University of Michigan Press: Ann Arbor, MI, 1966), pp. 158–85.

[6] Nash, J., 'Non-cooperative games', *Annals of Mathematics*, vol. 54 (1951), pp. 286–95.

[7] This game (VG) was originally suggested to Steven Brams by Danny Kleinman in 1983; the authors are pleased to acknowledge its source.

[8] Fishburn, P. C., 'Lexicographic orders, decision rules and utilities: a survey', *Management Science*, vol. 20 (1974), pp. 1442–71.

[9] Rapoport (note 5).

[10] Maschler (note 4).

[11] Brams, S. J. and Davis, M. D., 'The verification problem in arms control: a game-theoretic analysis', in Cioffi-Revilla, C., Merritt, R. L. and Zinnes, D. A. (Sage: London, forthcoming).

[12] Brams, S. J., *Superpower Games: Applying Game Theory to Superpower Conflict* (Yale University Press: New Haven, 1985), pp. 116–44.

[13] von Stackelberg, H., *Marktform und Gleichgewicht* (Julius Springer: Berlin, 1934).

[14] Rapoport (note 5).

[15] Brams, S. J. and Kilgour, D. M., 'Notes on arms-control verification: a game-theoretic analysis', in Avenhaus, R. and Huber, R. K., *Modelling and Analysis of Arms Control Problems* (Springer-Verlag: Berlin, forthcoming).

[16] Fichtner, J., 'Solution concepts for two person games which model the verification problem in arms control', in Avenhaus and Huber (note 15).

[17] Avenhaus, R., *Safeguard Systems Analysis* (Plenum: New York, 1986).

[18] The probabilities of error are $1-x$ and $1-y$; in statistics, an error of the latter kind (incorrectly signalling violation) is a type 1 error, and an error of the former kind (incorrectly signalling compliance) is a type 2 error).

[19] The authors ignore a fourth pure strategy for COL—consulting the detector and assuming it is incorrect by choosing A if it signals a violation and $\bar{A}$ otherwise. Unless x and y are small, this fourth strategy would simply complicate the analysis without adding any significant features to the game. In fact, the analysis focuses on relatively large values of x and y.

[20] Lynn, W. J., 'Existing U.S.–Soviet confidence-building measures', in Blechman, B. M. (ed.), *Preventing Nuclear War: A Realistic Approach* (Indiana University Press: Bloomington, IN, 1985), pp. 24–51, gives details on bilateral and multilateral agreements and assesses compliance with them.

Appendix 11A. A game-theoretic analysis

Maximin strategies and values

We begin by calculating the maximin strategies and values for the players in the Verification Game with detection, defined in section III, where we showed that the expected pay-offs of ROW and COL are

$$E_R(s; t, u) = s(u+ty)r_3 + (1-s)(t+u-tx) + (1-s)(1-t-u+tx)r_2 \quad \text{(A1)}$$

$$E_C(t, u; s) = s(u+ty) + s(1-u-ty)c_3 + (1-s)(1-t-u+tx)c_2 \quad \text{(A2)}$$

Observe that E_R can be rewritten as

$$E_R(s; u, t) = (1-s)r_2 + u[sr_3 + (1-s)(1-r_2)] + t[syr_3 + (1-x)(1-r_2)] \quad \text{(A3)}$$

Since the coefficients of u and t in (A3) are non-negative, it follows that

$$\min_{u, t} E_R(s; u, t) = E_R(s; 0, 0) = (1-s)r_2,$$

which is maximized by $s=0$. Thus, ROW's maximin strategy is $s=0$, and its maximin value is r_2. These maximin results are identical to those for VG without detection.

To find COL's maximin strategy, note from (A2) that

$$\frac{\partial E_C}{\partial s} = (c_3 - c_2) + u[1 - c_3 + c_2] + t[y(1-c_3) + (1-x)c_2] > 0. \quad \text{(A4)}$$

From (A4) it follows that

$$\min_s E_C(u, t; s) = E_C(u, t; 0) = (1-u-t+tx)c_2,$$

so COL's maximin value is c_2, and its maximin strategy is $u=0$, $t=0$ (or, if $x=1$, $u=0$, t arbitrary). Again, these maximin results are the same as for the no-detection case.

Inducement by COL

Inducement occurs when COL announces in advance its (mixed) strategy, inviting ROW to make its best response. In some detection games (Brams and Davis (note 11); Brams (note 12), ch. 4), COL can do better by means of this stratagem than by the players' simultaneously selecting maximin strategies.

Since ROW has only two pure strategies, COL can induce only one or the other (C or $\bar{C}$). Suppose that COL has chosen some u and t, and consider what ROW's best response is. From (A1) it follows that

$$\frac{\partial E_R}{\partial s} = -r_2 + u[r_2 + r_3 - 1] + t[yr_3 - (1-x)(1-r_2)]. \quad \text{(A5)}$$

Assume first that $x(1-r_2) + yr_3 \leq 1$. Then

$$yr_3 - (1-x)(1-r_2) = x(1-r_2) + yr_3 - 1 + r_2 \leq r_2,$$

so by (A5),

$$\frac{\partial E_R}{\partial s} \leq -r_2 + ur_2 + tr_2 = -r_2 + r_2(u+t) \leq 0$$

since $u+t \leq 1$.

We have shown that $s=0$ is ROW's best response to any strategy of COL if $x(1-r_2) + yr_3 \leq 1$. Thus, COL cannot induce $s=1$ in this case. This puts COL in an unfavourable position, for it is easy to verify from (A2) that COL cannot receive more than c_2 if $s=0$, and cannot receive less than $c_3 > c_2$ if $s=1$.

Now assume that (see figure 11.2)

$$x(1-r_2) + yr_3 > 1. \quad \text{(A6)}$$

Then $yr_3-(1-x)(1-r_2)>r_2$, and it is easy to show from (A5) that $\frac{\partial E_R}{\partial s}>0$, so COL induces $s=1$ iff (if and only if)

$$t>\frac{r_2+u(1-r_3-r_2)}{yr_3-(1-x)(1-r_2)}. \qquad (A7)$$

But $t\leq 1-u$ is also required, so u must satisfy

$$\frac{r_2+u(1-r_3-r_2)}{yr_3-(1-x)(1-r_2)}<1-u,$$

which can be proven to be equivalent to

$$u<\frac{x(1-r_2)-1+yr_3}{x(1-r_2)-r_3+yr_3}. \qquad (A8)$$

[From (A6), the right side of (A8) is well-defined; the numerator is positive and less than the denominator.] Hence, we have shown that, if (A6) holds, COL can induce $s=1$ by choosing any $u\geq 0$ satisfying (A8), and then any $t\leq 1-u$ satisfying (A7).

We now ask which, among the choices of u and t that induce $s=1$, does COL most prefer? This choice will make COL's inducement optimal. From (A3) it follows that

$$E_C(u, t; 1)=c_3+(1-c_3)(u+ty),$$

so COL maximizes its expected pay-off by choosing $(u+ty)$ as large as possible. If $y<1$, COL must choose

$$u=\left[\frac{x(1-r_2)-1+yr_3}{x(1-r_2)-r_3+yr_3}\right]^-; \quad t=1-u=\left[\frac{1-r_3}{x(1-r_2)-r_3+yr_3}\right]^+. \qquad (A9)$$

If $y=1$, any u satisfying (A8) and $t=1-u$ is sufficient.

We have shown that, if (A6) fails, COL cannot induce any strategy of ROW other than $s=0$; COL then obtains at most c_2. If (A5) holds, COL can induce either $s=0$ or $s=1$.

We next show that optimal inducement of $s=1$ is indeed beneficial to COL. Substitution of (A9) and $s=1$ into (A2) yields

$$E_C^*=E_C(u, t; 1)=\left[\frac{x(1-r_2)-(1-y)(1-c_3+c_3r_3)}{x(1-r_2)-(1-y)r_3}\right]^- \qquad (A10)$$

[The right side of (A10) is well-defined since the denominator exceeds the numerator, and the numerator exceeds $x(1-r_2)-(1-y)>x(1-r_2)+yr_3-1>0$, by (A6).]

Next we show that optimal inducement yields COL strictly more than c_3 by evaluating

$$x(1-r_2)-(1-y)(1-c_3+c_3r_3)-x(1-r_2)c_3-(1-y)r_3c_3$$
$$=x(1-r_2)(1-c_3)-(1-y)(1-c_3)$$
$$=(1-c_3)[x(1-r_2)-(1-y)]>0.$$

It is now clear that $c_3<E_C^*<1$.

Finally, we determine the effects of COL's optimal inducement on ROW by substituting (A9), and $s=1$, into (A1):

$$E_R^*=E_R(1; u, t)=r_3\left[\frac{x(1-r_2)-1+y}{x(1-r_2)-r_3+yr_3}\right]^- \qquad (A11)$$

[Again it is easy to use (A6) to show that the fraction on the right side of (A11) is well-defined.] To show that COL's optimal inducement yields ROW more than r_2, we calculate

$$x(1-r_2)r_3-(1-y)r_3-x(1-r_2)r_2+(1-y)r_2r_3$$
$$=(1-r_2)[x(r_3-r_2)-(1-y)r_3]>0$$

because, by (A6),

$$yr_3-(1-x)-xr_2>0,$$

so that

$$yr_3-r_3(1-x)-xr_2>0.$$

It now follows that $r_2 < E_R^* < r_3$.

Nash equilibria

We now determine all Nash equilibria of VG with detection. To avoid trivialities, assume that the detector satisfies $x>0$ and $y>0$.

We first identify all those equilibria with $s=1$. From (A2),

$$E_C(u, t; 1) = c_3 + (u+ty)(1-c_3),$$

so that at any equilibrium with $s=1$, COL must maximize $(u+ty)$ because $1-c_3 > 0$. Thus, for an equilibrium with $s=1$, either $u=1$, $t=0$, or if $y=1$, u is arbitrary and $t=1-u$.

To begin with, suppose $y<1$. If $u=1$, $t=0$, (A1) shows that

$$E_R(s; 1, 0) = 1 - s(1-r_3).$$

But since $1-r_3 > 0$, the choice of $s=1$ does not maximize ROW's expected pay-off; therefore, there are no equilibria with $s=1$ if $y<1$.

We next search for equilibria with $s=1$ under the assumption that $y=1$. If $u=1-t$, then (A1) shows that

$$E_R(s; 1-t, t) = 1 - tx(1-r_2) + s[tx(1-r_2) - (1-r_3)].$$

It follows that $s=1$ maximizes ROW's expected pay-off iff

$$t \geq \frac{1-r_3}{x(1-r_2)},$$

which, because $t \leq 1$ is required, can be arranged iff $x \geq \frac{1-r_3}{1-r_2}$. It can be verified that these necessary conditions are also sufficient.

We can now characterize all Nash equilibria with $s=1$ in VG:

I. *Co-operative equilibrium.* There is an equilibrium with $s=1$ iff

$x \geq \frac{1-r_3}{1-r_2}$ and $y=1$. In this case, these equilibria are precisely those combinations of strategies satisfying

$$s=1; \quad u=1-t, \quad t \geq \frac{1-r_3}{x(1-r_2)}.$$

Pay-offs at any equilibrium of type I are $(r_3, 1)$.

Analogously, we can identify all equilibria with $s=0$. First, from (A2),

$$E_C(u, t; 0) = [1-u-t(1-x)]c_2,$$

so COL maximizes its expected pay-off by picking $u=0$, $t=0$, or, if $x=1$, $u=0$, t arbitrary. If $x<1$, then by (A1),

$$E_R(s; 0, 0) = (1-s)r_2,$$

so $s=0$ is indeed a best response for ROW. If $x=1$, then (A1) shows that

$$E_R(s; 0, t) = r_2 + s[tyr_3 - r_2],$$

so $s=0$ maximizes ROW's expected pay-off iff

$$t \leq \frac{r_2}{yr_3}.$$

(This inequality holds for any choice of t if $y \leq r_2/r_3$; if $y > r_2/r_3$, some values of t are excluded.) Because the above necessary conditions can also be shown to be sufficient, we have

---

II. *Non-cooperative equilibria.* There are essentially two equilibria of this type. The strategy combination

$$s=0; \quad u=0, \quad t=0$$

is type IIa. There are no other equilibria with $s=0$ unless $x=1$, in which case the equilibria with $s=0$ are precisely those strategy combinations satisfying

$$s=0;\ u=0,\ t \leq \frac{r_2}{yr_3},$$

which is type IIb. At any type II equilibrium, pay-offs are (r_2, c_2).

Having identified all equilibria with $s=0$ or $s=1$, we turn to the case $0<s<1$. Differentiation of (A1) yields

$$\frac{\partial E_R}{\partial s} = -r_2 + u[r_2+r_3-1] + t[yr_3-(1-x)(1-r_2)]. \tag{A12}$$

Observe that $\frac{\partial E_R}{\partial s}=0$ at any equilibrium with $0<s<1$. It follows that there are no such equilibria with $t=0$, for, if $t=0$,

$$\frac{\partial E_R}{\partial s} = -r_2 + u[r_2+r_3-1] \leq \max\{-r_2,\ r_3-1\} < 0$$

by (A12). Now suppose that (see figure 11.2)

$$x(1-r_2)+yr_3 < 1.$$

Then

$$yr_3-(1-x)(1-r_2) = yr_3+x(1-r_2)-(1-r_2) < 1-(1-r_2) = r_2,$$

so that (A12) yields

$$\frac{\partial E_R}{\partial s} < -r_2 + ur_2 + tr_2 = -r_2 + (u+t)r_2 \leq 0$$

since $u+t \leq 1$. Again, no equilibrium with $0<s<1$ can exist.

Thus, to find all equilibria not of types I or II, we may assume that (see figure 11.2)

$$x(1-r_2)+yr_3 \geq 1; \tag{A13}$$

further, we need only consider strategies for COL with $t>0$. First, consider the case $u=0$, $t=1$. Substitution in (A12) shows that

$$-r_2 + [yr_3-(1-x)(1-r_2)] = 0,$$

which is equivalent to

$$x(1-r_2)+yr_3 = 1,$$

is necessary. From (A2),

$$E_C(u, t; s) = sc_3+(1-s)c_2+u[s(1-c_3)-(1-s)c_2]+t[sy(1-c_3)-(1-s)(1-x)c_2],$$

which we write as

$$E_C(u, t; s) = H + uK + tL, \tag{A14}$$

with H, K, and L defined appropriately. By calculus, the choice $u=0$, $t=1$ maximizes (A14) iff

$$L \geq 0,\ L \geq K.$$

The condition $L \geq 0$ is easily seen to be equivalent to

$$s \geq s_\ell(x, y) = \frac{(1-x)c_2}{y(1-c_3)+(1-x)c_2},$$

and the condition $L \geq K$ is equivalent to

$$s \leq s_u(x, y) = \frac{xc_2}{(1-y)(1-c_3)+xc_2}.$$

It is obvious that $0 \leq s_\ell < 1$, with equality iff $x=1$, and $0 < s_u \leq 1$, with equality iff $y=1$. Finally,

$$s_u - s_\ell = c_2(1-c_3)[x+y-1] > 0$$

since $x+y > x(1-r_2)+yr_3 = 1$.

It can be verified that these necessary conditions are also sufficient for an equilibrium with $u=0$, $t=1$. In summary, we have

III. *Constant-detection equilibrium.* There is an equilibrium with $0<s<1$, $u=0$, and $t=1$ iff $x(1-r_2)+yr_3=1$. In this case, such equilibria are precisely those strategy combinations satisfying

$$s_l \leq s \leq s_u; \; t=1, \; u=0.$$

Pay-offs at any equilibrium of type III are

$$(yr_3, \; xc_2+s[c_3+y(1-c_2)-xc_2]).$$

Next, under the assumption that (A13) holds, we search for equilibria with $0<t<1$ and $u=0$, and, of course, $0<s<1$. Substitution in (A12) shows that

$$-r_2+t[yr_3-(1-x)(1-r_2)]=0,$$

which is equivalent to

$$t=t_l(x,y)=\frac{r_2}{yr_3-(1-x)(1-r_2)}.$$

[As noted above, (A13) ensures that the denominator of this fraction is at least r_2 so that $0<t_l \leq 1$, with equality precisely when equality obtains in (A13).]

Now consider (A14). If the choice $0<t<1$ and $u=0$ maximizes COL's expected pay-off, it is necessary that $L=0$ and $K \leq 0$. As above, $L=0$ iff $s=s_l(x,y)$ and, assuming this to be true, $K \leq 0$ iff

$$-c_2+s_l[1-c_3+c_2] \leq 0.$$

This relationship can be shown to be equivalent to

$$c_2(1-c_3)[1-x-y] \leq 0,$$

which is certainly true, since, as noted earlier, (A13) ensures that $x+y>1$.

Because the above necessary conditions are also sufficient, we have

---

IV. *Never-accept (unconditionally) equilibrium.* There is an equilibrium with $0<s<1$, $u=0$, and $0<t<1$ iff $x(1-r_2)+yr_3>1$. In this case, such equilibria are precisely those strategy combinations satisfying

$$s=s_l; \; u=0, \; t=t_l.$$

Pay-offs at any equilibrium of type IV are

$$\left(\frac{yr_2r_3}{yr_3-(1-x)(1-r_2)}, \; \frac{yc_2(1-c_3)+(1-x)c_2c_3}{y(1-c_3)+(1-x)c_3}\right).$$

Analogously, we now assume (A13) and search for equilibria with $u=1-t$, $0<t<1$, and, of course, $0<s<1$. Substitution in (A12) and simplification shows that

$$t=t_u(x,y)=\frac{1-r_3}{x(1-r_2)-(1-y)r_3}$$

is a necessary condition for such an equilibrium. [It is easy to verify that $0<t_u \leq 1$, with equality precisely when equality holds in (A13).] The choice of $0<t<1$ and $u=1-t$ maximizes (A14) only if $L=K$ and $L \geq 0$. As observed earlier, $L=K$ iff $s=s_u(x,y)$. If $s=s_u$, $L \geq 0$ iff

$$-(1-x)c_2+s_u[y(1-c_3)+(1-x)c_2] \geq 0,$$

which is equivalent to

$$c_2(1-c_3)[x+y-1] \geq 0;$$

the latter inequality is a consequence of (A13).

Because these necessary conditions are also sufficient, we have

v. *Never-challenge (unconditionally) equilibrium.* There is an equilibrium with $0 \leq s < 1$, $u = 1-t$, $0 < t < 1$ iff $x(1-r_2) + yr_3 > 1$. In this case, such equilibria are precisely those strategy combinations satisfying

$s = s_u$; $u = 1 - t_u$, $t = t_u$.

Pay-offs at any equilibrium of type v are

$$\left(\frac{x(1-r_2)r_3 - (1-y)r_3}{x(1-r_2) - (1-y)r_3}, \frac{xc_2}{(1-y)(1-c_3) + xc_3} \right).$$

We now show that we have identified all Nash equilibria of VG with detection. If there is an equilibrium not already described, it can exist only when (A13) holds, and it must satisfy $0 < s < 1$, $0 < u < 1-t$, and $0 < t < 1$. Consideration of (A14) shows that a necessary condition for such an equilibrium is $L = 0$ and $K = 0$.

We have already noted that $L = 0$ iff $s = s_l$, and $L = K$ iff $s = s_u$. Since $s_u = s_l$ can be shown to be equivalent to

$c_2(1-c_3)[x+y-1] = 0,$

and since $x+y > 1$ is a consequence of (A13), we conclude that no such equilibrium can exist. Therefore, the only equilibria of VG with detection are those given by i–v.

Comparisons
When $x(1-r_2) + yr_3 < 1$, VG with detection is simple: ROW's strategy $s = 0$ (non-compliance) is dominant, and there is a unique (except when $x = 1$) equilibrium (IIa) with pay-offs (r_2, c_2). On the other hand, the situation is more complicated when $x(1-r_2) + yr_3 > 1$, for even with $x < 1$ and $y < 1$ there are two distinct new equilibria, IV and V, as well as the possibility of inducement by COL. We next compare these latter possibilities.

Assume that $x(1-r_2) + yr_3 > 1$, $x < 1$, and $y < 1$. First observe that $s_u > s_l$ by a calculation similar to the one above. Also $t_u > t_l$ iff $1 - r_2 - r_3 > 0$, and $t_u = t_l$ iff $1 - r_2 - r_3 = 0$. Finally, comparison of (A9) with the definition of t_u shows that COL's optimal inducement strategy is to consult its detector (D) just slightly more often than at the type v equilibrium, where the proportion is t_u.

We now compare the players' pay-offs at type IV and type V to show that V dominates IV, and that both dominate IIa. For COL, $E_C(\text{IV}) \geq c_2$, with equality iff $x = 1$, is obvious. The inequality $E_C(\text{V}) > E_C(\text{IV})$ is equivalent to

$$xc_2[y(1-c_3) + (1-x)c_2] > c_2[(1-y)(1-c_3) + xc_2][y(1-c_3) + (1-x)c_3],$$

which can be shown to hold iff

$x[y(1-c_2) + (1-x)c_2] > (1-y)[y(1-c_3) + (1-x)c_3].$

Now $x > 1 - y$ since (A13) holds, and

$[y(1-c_2) + (1-x)c_2] - [y(1-c_3) + (1-x)c_3] = (c_3 - c_2)(x+y-1) > 0,$

again because of (A13). This proves that (A15) holds, and we have $E_C(\text{V}) > E_C(\text{IV}) > E_C(\text{II}) = c_2$.

We now prove analogous inequalities for ROW's pay-offs at these three equilibria. First, it is easy to show that $E_R(\text{IV}) \geq E_R(\text{IIa}) = r_2$, with equality iff $x = 1$. The inequality $E_R(\text{V}) > E_R(\text{IV})$ can be shown to be equivalent to

$Q(y) = m - ny + py^2 > 0,$

where $m = (1-x)[1 - x(1-r_2)]$, $n = (1-x)(1+r_3) - xr_2$, and $p = r_3$. It is easy to verify that, if $\frac{1-r_3}{1-r_2} < x < 1$, $Q\left(\frac{1-x+xr_2}{r_3}\right) = 0$. The minimum of $Q(y)$ occurs at $y = \frac{n}{2p}$, and again, it can be checked that $\frac{n}{2p} < \frac{1-x+xr_2}{r_3}$, for $\frac{1-r_3}{1-r_2} < x < 1$. This completes the proof that $E_R(\text{V}) > E_R(\text{IV}) > E_R(\text{II})$ for $x(1-r_2) + yr_3 > 1$, $x < 1$, and $y < 1$.

Therefore, of the three equilibria that exist in the region $x(1-r_2) + yr_3 > 1$, $x < 1$, and $y < 1$, v is strictly preferred by both players. It is immediate that $E_R(\text{V}) < r_3$, and it can be shown that $E_C(\text{V}) \geq c_3$ iff $xc_2 \geq (1-y)c_3$.

We now compare the pay-offs at the dominant equilibrium, v, with the pay-offs under optimal inducement by COL. We have already noted that COL's strategies are only marginally different in these two situations. Comparison of $E_R(v)$ with (A11) shows that ROW receives slightly less under optimal inducement by COL than at v. [This also proves that $E_R(v)<r_3$.] Using (A10), it can be shown that COL's expected pay-off under optimal inducement, E_C^*, exceeds $E_C(v)$ exactly when

$$x[1-r_2-c_2+c_2r_3]+y[1-c_3+c_3r_3]>1-c_3+c_3r_3. \tag{A16}$$

It is easy to verify that the line defined by equality in (A16) lies below and to the left of $x(1-r_2)+yr_3=1$ (see figure 11.2). Therefore, the region where $E_C^*>E_C(v)$ is the entire triangle defined by $x(1-r_2)+yr_3>1$, $x<1$, and $y<1$, which is the shaded area shown in figure 11.2. Optimal inducement by COL thus increases COL's expected pay-off significantly over equilibrium v. Also, recall that $E_C^*>c_3$ whenever inducement by COL is possible.

Chapter 12. ARMCO-1: an expert system for nuclear arms control

ALLAN M. DIN

I. Formulation of the arms control problem

The average human mind has a distinct tendency to become overloaded when required to store more than a few precise numerical facts—perhaps as few as three or four. This means, for example, that most people probably feel quite lost in the continuing number game associated with the current NST (Nuclear and Space Talks) arms control negotiations in Geneva between the USA and the USSR; apart from the numerical fact of the day, one would, as a general rule, tend to forget both the basic, quantitative weapon balance of the superpowers as well as the previous arms reduction offers of either side!

It is of course commonly assumed that the minds of the NST negotiators are far above average (just how far and in which domain of knowledge may be open to question) and that, in any case, they have ready access to all the relevant data. Nevertheless, even expert negotiators are likely to have a hard time remembering, analysing and comparing all the numbers involved in nuclear arms control. A standard approach for overcoming this difficulty is to define more manageable subsets of the general problem which may be addressed sequentially or in parallel, but the risk is then that the overall perspective is lost.

Clearly, there is a lot more to nuclear arms control than the number aspect. Doctrines, perceptions (real and imagined), strategic assessments and psychology, for example, play a paramount role, but even if their role is crucial this should not be allowed to hide the fact that all arms control eventually has to come down to imposing both qualitative and quantitative constraints. The combination of all the above elements makes nuclear arms control highly complex, and more than warrants full recourse to both the art and science of negotiation.[1] Some prospects for demystifying the art and simultaneously developing the science of arms control negotiation are discussed below.

It is suggested that a novel ingredient in such an endeavour may be what has been coined Computer-Aided Arms Control (CAAC).[2] In the context of arms control negotiations, CAAC would mean using knowledge-engineering techniques and building expert systems[3] with the objective of making the computer an integral part in the analysis and decision-making process involved. It must be emphasized that the purpose is of course to make the computer a useful tool, but not a master, in the process and a tool which, by the very fact of having been developed, may be able to improve the analytical foundations of arms control.

Knowledge-engineering is (ideally) supposed to capture the essence of the expert knowledge of the experienced arms control negotiator and put it into a form comprehensible to a computer. That this can be done for the number game of arms control relating to the quantity and quality of weapon systems is obvious but, using artificial intelligence techniques and logic programming languages, it is also becoming increasingly feasible to address systematically the rule game which forms the backbone of the art of arms control negotiation.

The 'rule game' is used here as an euphemism for the interplay of doctrines, perceptions, and so on, which determines the proper negotiation platform and evaluates the arms control proposals of the other side. The concern of the knowledge-engineer is to find out precisely what constitutes the rule base used by the expert negotiator when making up his mind about the current issues. By interviewing a negotiator, the knowledge-engineer would, however, for several reasons most probably get no more than a very partial answer to his questions, at least initially.

A first reason could be that the expert may fear giving away sensitive information about negotiation positions. While it is true that details such as fall-back positions are among the most closely guarded secrets, such a fear would miss the point since the focus of interest is more the basic elements of analytical process than the results of the analysis. A second reason for the lack of response could be that the arms control expert is simply not able to formulate the rule base in a precise manner. It must be a basic premise, though, that there are some rules to be formulated, because otherwise arms control negotiations would have no prospects, logically speaking (cynics may of course claim that there are no prospects anyway). Supposing that the premise is correct, the question is then both 'how many' and 'which'.

The short history of knowledge-engineering has shown quite a few examples of areas of expert knowledge which were considered to be so complex as to merit the characterization of 'an art' but which, after closer scrutiny, turned out to be based on surprisingly few and simple rules. The story goes that a knowledge-engineer subjected an expert to questioning for a couple of days and that the latter eventually committed suicide when realizing that a lifetime of respected expertise amounted to no more than the application of a handful of rules of thumb. This should by no means be taken to suggest that negotiators of nuclear arms control will necessarily encounter a similar fate, but simply to propose probing whether or not something considered an expert art may eventually be broken down into logic building blocks.

Novel approaches to nuclear arms control ought to be particularly welcome in view of the impasse of the past years, which has given rise to much discussion about whether one is using a suitable approach and asking the right questions.[4] A number of analytical tools exist which are more or less relevant for problems of international security and the superpower relationship in particular, of which nuclear arms control is but one essential element. An interesting example is game theory, which can be used to treat a whole range of questions from the dynamics of warfare to the more static issues of verification and negotiation.[5]

There is of course already some understanding of the inner workings of decision-making processes for arms control in various countries and environments,[6] but an analytical, rather than descriptive, approach is needed to go beyond early attempts to simulate the relevant negotiation mechanisms.[7] System analysis and mathematical modelling techniques[8] offer many interesting possibilities for the quantification and demystification of arms control problems, but admittedly there is a long way to go before realistic applications are in view. Eventually however, such new approaches might be combined with earlier suggestions of non-standard and, hopefully, more effective frameworks for arms control negotiations.[9]

So far, the use of computers in problems of relevance to arms control analysis and negotiation has been very limited. Computer-aided decision-making[10] is a topic which has been developed over the years in areas such as economic development and crisis management, and the acquired experience could probably be put to good use in arms control problems. For a number of years, computers have been applied in strategic analysis, and in some new developments[11] a tendency is seen towards automated simulations in which 'intelligent' machines substitute for human players.

The following section describes an attempt to set up an expert system for nuclear arms control, ARMCO-1, with the basic purpose of illustrating how some of the problems and ideas discussed above might be taken into account in a pragmatic and evolutionary manner. Rather than calling ARMCO-1 an expert system, it would perhaps, at the present stage, be more appropriate to think of it as an example of logic programming containing some of the fundamental ingredients. As such, it might be able to motivate the further elaborate development process through which all expert systems must pass, from simple beginnings to fully-fledged tool. It should also be recalled that, in practice, an expert system never becomes perfect since there is always likely to be new knowledge to add.

ARMCO-1 focuses on nuclear arms control involving strategic and theatre nuclear weapons, because this is an arms control area of fundamental importance which has traditionally been treated somewhat in isolation. The success of an expert system is dependent upon a well delimited problem area. Nuclear arms control is admittedly quite a big problem area, but from the expert system point of view, and from a political point of view too, one may question whether it is big enough, that is, can it really be treated in isolation from other problems? The concluding section returns to the question of such issues as strategic defence (which is part of NST) and conventional arms.

In line with the expert system approach, it appears to be both reasonable and feasible to require ARMCO-1 to respond (more or less) intelligently to the following kind of query:

1. Data base queries, such as: 'How many warheads are there on Soviet strategic weapons with a range exceeding 7000 km?'
2. Queries on proposals for arms limitations, such as: 'The USA proposes to cut all land-based ICBMs by 50 per cent. What is the Soviet response? If negative, what are the reasons? Is there a counter-proposal?'

There is of course no unique way to set up an expert system capable of giving answers to the latter type of query, but there are several ways to test the ability of the system in order to approach a realistic performance. A first success would be attained if the computer, in response to the various nuclear arms control proposals put on the table by the superpowers during the past few years, could provide reactions similar to those reported officially. Clearly, the hope is to go even further.

II. Setting up an expert system

At the base of an expert system it is convenient to have an appropriate computer language, that is, an artificial-intelligence language such as Lisp or Prolog. This is not to say that it is impossible to use standard languages such as C, Ada and Basic, but it would be more complicated and time consuming. The basic difference is that the latter languages are tailored to tackle the algorithmic problems associated with the well known number-crunching capacity of computers, whereas the former are more suited for symbolic and logic manipulation.

The first language of artificial intelligence was Lisp (*List* processing). Even though it is still the foremost one it is often convenient to use a slightly higher-level language, Prolog (*Pro*gramming in *log*ic). This language has been adopted in the Japanese fifth-generation computer project and therefore symbolizes to some extent the high hopes attached to future practical applications of artificial intelligence.

For quite some time, logic programming was the preserve of computer specialists working on large mainframe machines. This has been one of the principal obstacles to wider use of artificial intelligence, since most people working in domains of potential practical applications were unlikely to have any contact with, let alone any knowledge about, the higher spheres of computer science. However, with the advent of the personal computer, the gap between the digital machine and the end user is continuously narrowing. Still, the number of people knowing something about both artificial intelligence techniques *and* any given, potential application area is extremely small. The hope is of course that the existing gaps may be overcome by knowledge-engineering.

The programming of ARMCO-1 has been done using a version of Prolog called Micro-Prolog[12] in an adaption for the Apple Macintosh computer.[13] This is more than adequate for setting up a small expert system where the problem is just to develop the basic ideas and to demonstrate the principle. If deemed successful, it will of course be possible to transfer developments of the expert system to a high-performance machine when the data and rule base grow too big.

The basic data about strategic and theatre nuclear weapons included in ARMCO-1 are adapted from the *SIPRI Yearbook 1986*.[14] In the (simple) syntax of Prolog, the data base will thus contain statements such as:

US-ICBM(Minuteman-III)
SU-SLBM(SS-N-20).

These are so-called prefix sentences for unary relations which translate to the natural-language form (using the abbreviations US for the United States, SU for the Soviet Union, ICBM for intercontinental ballistic missile and SLBM for submarine-launched ballistic missile):

Minuteman-III is a US ICBM
SS-N-20 is a SU SLBM.

In Prolog the relation (or property) name 'US-ICBM' is written as one word (using a hyphen), and the same applies to the object 'Minuteman-III', which has the property. A more elaborate expert system may work entirely with natural-language sentences, which are parsed into an equivalent Prolog syntax, but this is not an essential point in the present context. The main thing is to set up a characterization (like the statements in the first example above) for all the relevant weapon systems in terms of their properties: ICBM, SLBM, bomber, land-aircraft, land-missile, naval-aircraft, naval-missile and their affiliation, US or SU.

A further step is to define a dictionary of terms to be used in the dialogue with the expert system. For example, the computer should know that if a weapon has the property of being a US-bomber, then it also has the property of being a US-strategic-weapon:

US-strategic-weapon(_x) if US-bomber(_x).

Here _x is a variable name which may be any of the weapons that ARMCO-1 has been told about. The above statement is characteristic of expert systems based on so-called production rules, that is, 'if . . . then . . .' type of sentences; Prolog can of course handle much more complex sentences. The dictionary can be extended to include any of the commonly used terms (weapon-system, strategic-weapon, theatre-weapon, etc.) which feature in the nuclear arms control vocabulary.

The data base includes information about various characteristics of the nuclear weapons; a statement about their number, for example, would be represented as:

Number-of(SS-4 112).

This is a binary relation with the relation name 'number-of' and the two arguments 'SS-4' and '112', which in natural language amounts to:

The number of SS-4s is 112.

The number here refers to the number of launchers. Similarly, there is information about the number of warheads on each launcher, which is expressed as the binary relation:

Warheads-of(Minuteman-III 3).

Furthermore ARMCO-1 contains data about the ranges of the weapons, for example the statement:

Range-of(Backfire 3700),

which gives the information that the Soviet land-aircraft 'Backfire' has a range of 3700 km. Finally, there are statements about the deployment years of the various weapons, for example:

Year-of(Trident-1 1979).

Additional data concerning warhead yields, for example, and a more detailed breakdown of the weapon types and their characteristics than presented above, may be accommodated without difficulty. One of the virtues of logic programming (at least in principle) is precisely that new data and knowledge can be added to the expert system when the need arises without interfering with its inference mechanism.

With the input described above, ARMCO-1 is already capable of conducting an 'intelligent' dialogue concerning complex queries of the data base. For example, one might want to ask the question: 'Which US multiple-warhead strategic weapons with a range above 5000 km were deployed in the 1970s?' In principle it would be possible to let the expert system parse the above query into the more rudimentary Prolog syntax (in practice this might still be beyond the state of the art):

US-strategic-weapon(_x) & warheads-of(_x_y) & LESS(1_y) & range-of (_x_z) & LESS(5000_z) & year-of(_x_v) & LESS(1969_v) and LESS(_v 1980).

Here _x is the variable name for the weapons that are being queried, and the expert system program simply scans through its data base to find the types which fulfil the requirements on warheads, range and year of deployment. ARMCO-1 rapidly produces the answer:

Minuteman-III, Poseidon, Trident-I.

This feature is not very impressive in itself but illustrates how an expert system works in a simple situation, providing intelligible answers to naturally sounding questions.

After having established an appropriate knowledge base involving numerical data and a dictionary for the relevant terminology, the subsequent (and much more difficult) problem is how to construct a rule base which in combination with the inference mechanism of Prolog is able to provide 'intelligent' answers to the type of query concerning arms control proposals mentioned in the first section. A computer running an expert system 'reasons' by way of built-in techniques (such as forward and backward chaining) and under the guidance of the particular rule set-up; this part of the expert system is sometimes referred to as the 'inference engine'.

It is a matter of convention whether the rule base is considered to belong to the knowledge base or to the inference engine. On the one hand, the rules governing arms control analysis may be said to be part of the knowledge base providing the foundation of the reasoning; on the other hand, the way the rules are translated and applied (for example in which order) very much determines the mechanisms and the paths of inference. In any case, for the expert system to be effective, the inference engine has to be structured in such a way as to prevent the occurrence of free-wheeling search processes which would either result in too many answers or in the computer simply running out of memory (or otherwise producing errors).

A well-structured inference engine often warrants an algorithmic approach, or other explicit control structures, to the search for answers thus running counter to some common credos about logic programming and expert systems. Although an expert system with few rules may not explicitly appear to use any algorithm or operate sequentially, one is nevertheless clearly able to distinguish an implicit control structure which guides the flow of reasoning. When the expert system grows bigger, the overview of the control structure becomes blurred because the search space 'explodes', that is, the number of possible cases which must be investigated grows exponentially. A way out of the 'exploding' search-space problem is to use heuristics, that is, some kind of rules of thumb and/or evaluation functions whereby an algorithm provides a numerical value indicator which allows definite (and hopefully best possible) choices to be made.

A simple way of applying the evaluation function approach to ARMCO-1 is to assign values to the various nuclear weapons on a scale from 1 to 10 depending on the different characteristics. For example, it may be reasonable to ascribe a quite high value, say 9, to the more or less invulnerable SLBMs with their long range of, say, more than 5000 km. In Prolog syntax this would be expressed as:

Value of (_x 9) if SLBM(_x) & range-of(_x_y) & LESS(5000_y).

The above statement will assign the value 9 to all weapons (with variable name _x) fulfilling the condition. It would seem reasonable to count this value for each warhead of the particular weapon launcher. Similarly, the other weapons may be assigned values according to their importance with, say, older, short-range theatre weapons appearing at the lower end of the scale. More complicated schemes may of course be considered in which, for example, different scales are assigned to the strategic and theatre weapon types. In the simplest case, however, there will just be one evaluation function which is the sum of the values of all categories of weapon.

The value-assignment exercise described above is invariably somewhat arbitrary. For example, in one scheme considered, the USA came out with an approximate total strategic weapon value of 83 000 and a theatre weapon value of 31 000; the Soviet Union had total values of 72 000 and 40 000, respectively. Although oversimplified, such an example may be considered sufficiently interesting to develop the expert system further along these lines provided one agrees to the implied force relationship of an overall balance, with a Soviet edge in theatre weapons and a US edge in strategic weapons (if not, then other assessments may be considered as a starting point).

Following this kind of scheme (with whatever agreed values), acceptable arms reduction proposals would have to conform to maintaining an approximate value parity between the two sides. If, for example, a proposal were put on the table that both sides reduce the number of ICBMs by 50 per cent, then the expert system would proceed to evaluate the best way to distribute the cuts among the selected weapon types so as to optimize the resulting total value function on both sides. In this case, the Soviet Union would no doubt reject the proposal since the Soviet value function would be diminished more strongly than the US one. The expert system could then be made to explain that the reason for this negative response is that the Soviet Union has proportionally more ICBMs than the USA. Subsequently, a counterproposal could be presented using the same technique.

The above exercise is simple, but it illustrates the effectiveness of value functions in limiting the search. With a more sophisticated value assignment, the proposal analysis could certainly be made somewhat more realistic; nevertheless, the expert system should also contain rules more oriented towards qualitative and heuristic reasoning. These could, for example, express the reluctance to dismantle newer weapon systems or draw on some historical experience of the reactions of one side or the other to particular types of reduction. There are likely to be many other relevant rules which may be both difficult to extract and to formulate, and focusing attention on them in an expert system perspective could hopefully give a positive contribution to making the arms control analysis more transparent.

III. Conclusions and outlook

The current interest in expert systems for the most diverse domains of knowledge may turn out to be rewarding in many respects. Obviously, if an expert system is achieved for a particular problem area and its ability in emulating knowledge is proved through practical applications, then the experience may be deemed successful. But even if this is not the case, something may have been gained in the process. For example, setting up an expert system requires the problem at hand to be properly defined and delimited; if this turns out to be difficult or impossible, the conclusion could be that the problem has no solution and should be redefined or enlarged.

Another possible positive outcome is a clear separation of the qualitative from the quantitative aspects and a better understanding of the problem in terms of its implicit and explicit elements—all preconditions for any rational approach to problem-solving. In the nuclear arms control context there has been little progress along these lines, as far as is seen from past public statements. There are, however, some signals that the two superpowers are about to leave the somewhat sterile number game behind and enter the more substantial rule game. Here, the issue is to define, and possibly formalize and/ or quantify, measures for national and international security.

Such as analysis must incorporate a rethinking of the foundations of nuclear deterrence[15] as well as its future in a world in which other doctrines may develop. As a consequence, it might become generally accepted that making progress in nuclear arms control—with more than an incremental perspective in mind—requires taking account of the possible future role of strategic defence, and in fact any kind of alternative defence ideas, as well as the level and structure of conventional forces. But although nuclear arms control cannot really be discussed in isolation, the current NST approach (as well as the expert-system approach discussed above) is not unwarranted; substantial nuclear arms reductions are still desirable and possible in the near-term perspective, disregarding other problems, simply because the stockpiles are so big.

Eventually, however, the problem of nuclear arms control must be treated in its wider context to include strategic defence and conventional weapons. This is not likely to make arms control negotiations easier, but at least the problem may be well posed and well delimited. While the latter is a necessary condition for the successful use of expert-system techniques in analysing the long-term perspectives of international security by modelling and simulation, there is of course no guarantee that it would also be sufficient.

Notes and references

[1] Raiffa, H., *The Art and Science of Negotiation* (Harvard University Press: Cambridge, MA, 1982).

[2] Din, A. M., 'Strategic computing', SIPRI, *World Armaments and Disarmament: SIPRI Yearbook 1986* (Oxford University Press: Oxford, 1986).

[3] Hayes-Roth, F., Waterman, D. A. and Lenat, D. B., *Building Expert Systems* (Addison-Wesley: Reading, MA, 1983).

[4] Schelling, T. C., 'What went wrong with arms control?', *Foreign Affairs*, vol. 64, no. 2 (Winter 1985/86), p. 219; Kubbig, B. W., '(Re-)Defining and refining the criteria for nuclear arms control', *Bulletin of Peace Proposals*, vol. 16, no. 3 (1985), p. 199.

[5] Brewer, G. D. and Shubik, M., *The War Game* (Harvard University Press: Cambridge, MA, 1979); Axelrod, R., *The Evolution of Cooperation* (Basic Books: New York, 1984); Brams, S. J., *Superpower Games* (Yale University Press: New Haven, CT, 1985).

[6] Brauch, H. G. and Clarke, D. L. (eds), *Decisionmaking for Arms Limitation* (Ballinger: Cambridge, MA, 1983).

[7] Bonham, G. H., 'Simulating international disarmament negotiations', *Journal of Conflict Resolution*, vol. 15, no. 3 (1971), p. 299.

[8] Avenhaus, R. et al., 'Systems analysis and mathematical modelling in arms control', *OR Spektrum*, vol. 8 (1986), p. 129.

[9] Calogero, F., 'A scenario for effective SALT negotiations', *Science and Public Affairs*, June 1973, p. 16.

[10] Chacko, G. K., *Computer-Aided Decision-Making* (American Elsevier Publishing Company: New York, 1972).

[11] Bennett, B. W. and Davis, P. K., 'The role of automated war gaming in strategic analysis', Rand Corporation, Dec. 1984.

[12] Clark, K. L. and McCabe, F. G., *Micro-PROLOG* (Prentice-Hall: Englewood Cliffs, NJ, 1983).

[13] LPA MacPROLOG, Logic Programming Associates, London, 1986.

[14] Arkin, W. M. et al., 'Nuclear weapons', SIPRI, *World Armaments and Disarmament: SIPRI Yearbook 1986* (Oxford University Press: Oxford, 1986), p. 37.

[15] Powell, R., 'The theoretical foundations of strategic nuclear deterrence', *Political Science Quarterly*, vol. 100, no. 1 (Spring 1985), p. 75.

Index

Abe 98
ACRONYM system 170
Ada 11, 18
Advanced Landsat Sensor 171
AI (artificial intelligence):
 applications 40–1
 architecture 47
 arms control and 8, 20–3, 26, 81
 automated tactical battlefield and 100–19
 C³I and 25
 civilian 5–6
 defining 3–5, 33–4
 emphasis changes 41
 exaggerated claims for 24, 45, 158
 failures of 77
 game theory 22
 hardware 12, 24, 45
 heuristic search 12, 37
 human control and 25, 159
 integrated battlefield and 121–2
 knowledge representation 37–40
 languages 12, 13, 34–5, 48–51
 limitations 147–8, 157
 memory space 71
 military applications 14–20, 26, 76–8, 106–7, 117–19
 problems 34 *see also* limitations
 programming 34, 35
 prospects 45–6
 reliability questioned 45
 research funding 76
 scene description and 169–71
 search 35–7
 search space 12
 software 12, 24, 25, 47
 software writing by 148–9
 specificity 45
 strategic stability and 160
 techniques 35–40
 time and 26
 weapons projects 15–16
 see also automated tactical battlefield; computers; expert systems
Airborne Command and Control System 105
AirLand Battle doctrine:
 AI and 8, 18, 108
 automation and 102
 command and control 17
 expert systems and 16, 103
 high technology and 101
 nuclear weapons and 120, 123, 124
 offensiveness 126, 127
 SCI and 96, 97

weapons integration 102
algorithms 12, 13, 47, 48, 172
All Source Analysis System 104
America *see* United States of America
ARMCO–1 214–22
arms control:
 modelling 21–2
 simulation 21–2
 see also following entries
arms control negotiations 22, 214–17
arms control verification:
 computers and 8, 20–3, 165–78
 game analysis 193–213
 satellites 8, 20, 21
Army Data Distribution System 105
artificial intelligence *see* AI
ASAT (anti-satellite activities): BMD and 135
automated tactical battlefield:
 AI and 100–19
 current efforts 104–5
 distributed knowledge-based systems 112
 history of 100, 101
 nuclear weapons control 119–23
 political control 129–31
 problems of 105–6
 resistance to 109
 sensor fusion 109
 simulation systems 112–13
 weapon fault diagnosis 113
autonomous land vehicle 15, 44, 92, 93–4, 96, 98, 117
Autonomous Smart Weapons Program 98

Basic 10, 34, 82
Battlefield Exploitation and Target Acquisition system 110
battlefield robotics 15, 26, 117
BC-machine 65–6
Bekaa Valley 158
BiMOS 69
'blackboard architecture' 51
BMD (ballistic missile defence):
 analysis and design 137–40
 ASAT and 135
 cost 136
 deterrence and 149–50
 escalation and 146–7
 failures 143–5
 reliability 140–2, 149
 sabotage 144–5
 software and systems 136–43, 148

INDEX

C (language) 50, 55, 59
C³I:
  AI and 18
  automation 104–5, 107–12
  definitions 182
  functions 157
  nature 16–18
  survivability 122
caches 52–3, 57, 58
Caltech 65
Carl Vinson 16
circuit testing 61
CMOS technology 57, 68, 69
Cobol 10
coherent optics/holography 166
Common Lisp 50
Compact Lisp Machine (CLM) 57, 58
comprehensive test ban 20, 142, 150
Computer Aided Arms Control 214
computers:
  architectures 9, 72, 173–4
  arms control negotiations and 22
  arms control verification and 165–78
  BMD and 136, 137
  chips 6, 9, 19
  decision-making and 5, 7
  development of 33
  failures in 24
  fifth generation 3
  hardware 8–10
  history 8–9
  image processing 166–9
  languages 3, 10–11, 71
  man–machine interface 145
  military using 14–15
  miniaturization 14
  natural language and 5
  negative aspects 25
  optical 70
  parallelism 172–3 *see also* processing, parallel
  pervasiveness of 7
  sabotage 144
  sensor data and 20
  software 8, 10–11, 33, 45, 148–9
  user interface 9–10
  vision 166, 170
  weapon connection 6–8
  see also AI; expert systems; systems
computing: symbolic versus numerical 47–8
confidence-building measures 161
Connection Machine 66–7
Cosmic Cube 65
Council of Mutual Economic Assistance 153
counter-air 103, 109, 127
crisis instability 26
crisis stability 126–9
crossbar topology 62

DADO machine 67

DARPA (Defence Advanced Research Projects Agency) 7, 15, 88–90, 95, 110
'data' 167
data compression 171
data transformations 167
deep strike 102, 110, 124, 126, 127, 128
DENDRAL 43
deterrence *see* nuclear deterrence
Digital Equipment Corporation 44
digital signal processors (DSPs) 67
disarmament *see* arms control
Distributed C³I project 112

Eastport panel 141
EEC (European Economic Community) 153
electronic warfare 77, 118
EMP (electromagnetic pulse) 9, 14, 16, 122
encryption 14
Enemy Situation Correlation Element 104
ENIAC 88
escalation:
  control of 179, 181, 191
  military doctrines and 123–30
ESECS 127
ESPIRIT 153
Europe, war in 103
expert systems:
  architecture 42
  arms control and 20, 22, 214
  building 42, 44, 217–21
  C³I 158
  characteristics 170
  computer vision and 170
  default values 148
  definition 147
  examples of 43–4
  frames 148
  image understanding 170
  languages 13, 71
  limitations 44
  military applications 44
  nature of 13, 41, 50–1
  nuclear weapons control 121
  parallelism 63–4
  problems of 44–5
  production rules 148, 157, 170
  tactical command decision aids 108–9, 118
  time and 25, 26
  value doubtful 158
  weakness 148
  weapon fault diagnosis 113
  working of 42–3
  see also AI; computers
Explorer 57, 58

Fast Fourier Transforms (FFTs) 67, 172
Field Manual FM 100–5 123
first strike 126, 127
FOFA (follow on forces attack) 17, 101, 102, 104, 111, 124, 125, 126

Force Agreement 187
Fortran 10, 55
frame-based systems 39

gallium arsenide 9, 68–9, 72, 172
game theory 22, 179, 207–13, 216
gaming 21, 184–5, 186, 194–205, 207–13
garbage collection 55
Geneva arms control negotiations 214
Germany, Federal Republic 124
global bus 64, 65
granularity 62–3

HEROS system 17
heuristics 75, 78, 185–6 *see also* AI: heuristic search
Hughes Missile Systems Group 115
hydrogen bomb 7, 23

'image' 167
image analysis 174–7
image interpretation 115–17, 177–8
image parallelism 173
image processing 115, 166–9, 171–4
imagery data 168
'imitation game' 4
instruction 'prefetch' 52–3
integrated battlefield 120–3
Intel Corporation 65
'intellectoid' 98–9
intelligence: human and machine 75–82
interconnection networks (ICN) 61–2
Interlisp 58
Israel 158

Japan 5, 6
Japanese fifth generation computers 5, 87, 89, 217
Joint Surveillance Target Attack Radar 18
Joint Tactical Fusion Program 104
Joint Tactical Information Distribution System 104–5
Josephson devices 172

Landsat system 171
lasers 136
'launch on warning' 144
Libya 158
linguistic analysis 81
Lisp language 3, 12, 35, 49–50, 52, 71, 217
Lisp machines 56–8

machine vision 115–17
Maneuver Control System 104
mapping imaging 170
matrix transposition 172
memory architecture 60–1
microcode 56
microelectronics 100, 101
Micro-Prolog 217

military doctrines:
  escalation risks and 123
  technology and 101–2
  see also AirLand Battle: FOFA
mil-spec computers 14
MIMD (Multiple Instruction Multiple Data) 65, 173
missiles, ballistic 103
missiles, cruise 125, 160
missiles, guided 76, 105, 115
MIT 56, 57, 76
modelling 21–2, 78
models 179
Modula 11
MOS technology 69
Motorola M68020 68
MSI (medium-scale integration) 56
Multibus I 58
Multibus Interface Module 57
Multispectral Target Acquisition System 105
MYCIN 43

NASA 88
Nash equilibria 194, 201–4
national technical means 165
National Test Bed 19
NATO (North Atlantic Treaty Organization):
  ACCIS (Automated Command and Control Information System) 17
  C^3I 17, 18
  first use doctrine 124
  flexible response 120, 124
  military doctrines 17, 101, 127, 129, 160
  see also Airland Battle; FOFA
  nuclear weapons control 119
navy battle-management programme 16, 44, 92, 96, 147
Neumann, John von 8, 9, 89
Ninja programme 114–15
NORAD 144
NuBus 57, 58
nuclear deterrence 7, 179, 181, 191
nuclear strategy 179–91
nuclear threshold 123, 124–6
nuclear war 78, 119, 120, 121, 123
nuclear weapons: advent of 7, 155
nuclear weapons, tactical: command and control 119–23

optical computers 70
'Order of Battle Version 1 Knowledge Based' system 110

packet switching 61
parallel machines 64–7
Pascal 11, 34, 55, 59
pattern recognition 168
Pentomic division 120
Pershing II missile 125

Personal Scientific Computer 65
Personal Sequential Inference (PSI) machines 58–9
Phoenix missile 142
pilot's associate 15–16, 44, 92, 96, 113–14
pipelining 53–4, 55, 173, 177
pointers, tagged 54
Position Location Recording System 104–5
precision-guided sub-munitions 124
Prisoner's Dilemma 80, 81–2
processing:
   neighbourhood 173
   parallel 6, 9, 16, 45, 49, 60–7, 71
   sequential 56–9
   speed 6
   vector 173
programming:
   automatic 11, 148
   exploratory 35, 47
   logic 24–5
   structured 11
Prolog language 3, 12, 35, 48–9, 54, 71
Prolog machines 58–9
PROSPECTOR 43

R1 system 44, 45
Rand-Abel 90
RAND Corporation 109, 112–13, 114, 180, 186, 187
Rapoport, A. 193
Reduced Instruction Set Computer 70–1
Rete algorithm 51
Revolutionary New Generation System Tool 98
robots 4, 5, 26, 41, 115 *see also* battlefield robotics
rules of engagement 138, 142
runway-cratering sub-munitions 103
Russia *see* Union of Soviet Socialist Republics

SACRA system 17
SALT agreement 165
satellites:
   on-board versus ground processing 171–2
   verification by 165, 171–2
scene description 168–71
SCI (Strategic Computing Initiative):
   achievement objectives 94
   functional objectives 94
   funding 95, 98
   overview 90, 91
   purpose 7, 44, 99, 107
   second programme 98
SDI (Strategic Defense Initiative): 8, 18–19, 24, 88, 140, 154
semantic network 37–8
shared memory machines 64
signal processing 41, 45, 67
signal theory 166

signal understanding 168
SIMD (Single Instruction Multiple Data) 64, 66, 173, 174, 177
simulation 21–2, 142, 179–91, 182–4
SIOP (Single Integrated Operational Plan) 17
'smart' weapons 14, 101, 158
space mines 140
space shuttle 25, 141
speech recognition 40, 45
SPOT satellite 21
SSI (small-scale integration) 56
stacks 54–5
Strategic Computing document 90, 92
strategic defence: software and systems 135–50 *see also* BMD
supercomputers 6
symbol processing 35, 51–6, 71, 72
synthetic aperture radar 20, 167
System Dynamics 184
systems:
   evolution 143
   failures 143–5
   integration 142

Tactical Air Target Recommender 109
tactical command decision aids 108–9, 118
tanks 105, 114–15, 117
Target Analysis and Planning system 122
technological progress: stability and 154–6
TEMPEST computers 14
TERCOM guidance systems 160
thought, human 4, 34, 75
'time bombs' 144, 145
tomographic data 167
topologies 62
'trap doors' 144–5
TRW Defense System Group 110
TTL (transistor-transistor logic) 56
Turing, A. 4
TWIRL 112

unification operation 55
Union of Soviet Socialist Republics:
   computerization 14
   crisis behaviour 121
uni-processors 71
United States of America:
   Defense Department 18, 69, 93
   Defense Nuclear Agency 122
   National Command Authority 123, 124, 147
   UTACC (USAREUR Tactical Command and Control System) 18

verification game 194–205
Very Intelligent Surveillance and Target Acquisition (VISTA) project 110, 111
VHSIC (Very High Speed Integrated Circuits) 9, 19, 69–71, 119

Viet Nam War 101
virtual memory 51–2
VLSI (Very Large Scale Integration) 9, 19, 56, 57, 66, 67, 68–71, 72

war, accidental 8, 26, 150
war-gaming 21
Warren Abstract Machine (WAM) 59
war termination 179, 181, 191
WASP anti-armour missile project 115
WAVELL system 17

weapon fault diagnosis 113
Wide Area-armour Munitions (WAAM) programme 115
World War II 6
WTO (Warsaw Treaty Organization): computerization 14
WWMCCS (Worldwide Military Command and Control System) 17

Xerox Lisp machines 58